D0713213

# Early Earthquakes of the Americas

There is emerging interest amongst researchers from various subject areas in understanding the interplay of earthquake and volcanic occurrences, archaeology and history. This discipline has become known as archaeoseismology. Ancient earthquakes often leave their mark in the myths, legends, and literary accounts of ancient peoples, the stratigraphy of their historical sites, and the structural integrity of their constructions. Such information leads to a better understanding of the irregularities in the time–space patterns of earthquake and volcanic occurrences, and whether they could have contributed to some of the enigmatic catastrophes in ancient times.

This book focuses on the historical earthquakes of North and South America, and describes the effects those earthquakes have had with illustrated examples of recent structural damage at archaeological sites. It is written at a level that will appeal to a wide variety of individuals with different academic backgrounds. Students and researchers in the fields of earth science, archaeology, and history will greatly benefit from this book.

ROBERT KOVACH is Professor of Geophysics and Associate Chairman of the Department of Geophysics at Stanford University. He has conducted geophysical research in Mexico, Pakistan, and the Middle East and is actively engaged in the cross-fertilization between geophysics and archaeology. He is the author of more than 200 scientific publications. Professor Kovach was the recipient of a John Guggenheim Fellowship in 1971 and was Invited Professor, Japan Society for the Promotion of Science in 1975. He designed seismic experiments for the Apollo 14, 16, and 17 missions to the Moon, for which he received the National Aeronautics Space Administration Medal for Exceptional Scientific Achievement. He is a past-president of the seismology section of the American Geophysical Union, a fellow of the Geological Society of America, and a member of the Seismological Society of America and the Society of Exploration Geophysicists. Professor Kovach is also the author of the successful text *Earth's Fury: An Introduction to Natural Hazards and Disasters* (1995).

Crushed skeleton of a man caught beneath a collapsed wall. Reproduced courtesy of the Peabody Museum of Archaeology and Ethnology, Harvard University.

# Early Earthquakes
# of the Americas

PROFESSOR ROBERT L. KOVACH

Department of Geophysics,
Stanford University

 CAMBRIDGE
UNIVERSITY PRESS

PUBLISHED BY THE PRESS SYNDICATE OF THE UNIVERSITY OF CAMBRIDGE
The Pitt Building, Trumpington Street, Cambridge, United Kingdom

CAMBRIDGE UNIVERSITY PRESS
The Edinburgh Building, Cambridge, CB2 2RU, UK
40 West 20th Street, New York, NY 10011-4211, USA
477 Williamstown Road, Port Melbourne, VIC 3207, Australia
Ruiz de Alarcón 13, 28014 Madrid, Spain
Dock House, The Waterfront, Cape Town 8001, South Africa

http://www.cambridge.org



First published 2004

Printed in the United Kingdom at the University Press, Cambridge

*Typeface* Swift 9.5/14 pt. *System* LaTeX 2ε [TB]

*A catalogue record for this book is available from the British Library*

ISBN 0 521 82489 3 hardback

# Contents

*The plates are situated between pages 84 and 85.*

# *Preface*

This book involves a number of disciplines, but its foundation is seismology, the study of earthquakes. In recent decades, the science of seismology, in particular the study of individual earthquakes has expanded dramatically. Its role and importance in understanding how ancient societies and cultural centers were developed and abandoned, however, has not received as much attention. One would like to be able to simply open a window of history of past earthquake occurrences by referring to a written record, but there are many temporal and spatial gaps, omissions, and shortcomings in such a record prior to the nineteenth century. The written record for earthquakes only takes us back as far as the fifteenth century in the Valley of Mexico. Even if the written record were complete for the Americas we are still faced with a short historical time span. Considering our knowledge today of the geography of earthquakes in the Americas it is difficult to believe that earthquake occurrences did not take place prior to this time.

Early written records are useful but often reflect biases and religious beliefs of their scribes together with human exaggerations. As a result, we are left with the problem of factoring and interpreting truth from oracular propaganda. It is here that we can turn to the archaeological record. Just as many archaeologists would argue that pottery sherds speak to them, a seismologist can look for evidence of past earthquake occurrences in the material remains eagerly excavated by archaeologists. We might as a result uncover evidence by *dirt seismology* of past events, which have occurred and were not recorded or overlooked in historical texts.

For a number of years I have led a collegial dialogue at Stanford University entitled "Earthquakes of the Americas." It started as a discussion group to explore a number of interdisciplinary questions with earthquakes as an underlying theme. How did early native cultures such as the American Indians, the Aztecs, and the Maya perceive earthquakes and other natural disasters? Can we extract facts from early myths, legends, petrographs, and glyphs? Is there evidence for past earthquake occurrences in the archaeological sites of the Americas? Could earthquakes have played a role in the abandonment of ancient settlements? In the past were earthquakes used to justify certain clerical or political viewpoints? As the dialogue evolved so did the number of earthquakes examined, providing fuel for a fascinating intellectual trail through the areas of archaeology and anthropology. The earthquake hypothesis has, of course, its critics. Evidence is often patchy and circumstantial. Yet the examination of the seismic evidence when coupled with the human historical record may eventually prove that earthquakes have been a major contributing factor towards cultural collapse.

A vast field of literature exists describing the archaeological expeditions of the Americas beginning with the explorations in the latter half of the nineteenth century carrying through the twentieth century. However, since the unifying theory of plate tectonics had not been fully developed until the late 1960s most archaeologists could not fully appreciate its implications and ramifications as the driving force for the temporal and spatial pattern of earthquake occurrences. As a result the possibility of ancient earthquake damage was generally overlooked as various archaeological sites were discovered, excavated and reconstructed to their *believed* original state.

In examining the earthquake history of the Americas it is evident that not every major earthquake of the past has left an archaeological trail of destruction. However, these earthquakes may have left geological and geomorphic manifestations. Rock horizons that have been horizontally offset or vertically uplifted can often be placed in a historical time framework. Seismically induced liquefaction has left evidence for sand expulsion features that can be recognized in coastal plain regions and dated to estimate ages of prehistoric earthquake episodes. These investigations form the basis of *paleoseismology*.

A motivation for this book was the realization that there was no single book focused on the Americas that examined the interplay and implications between archaeology, myths and legends, and past earthquake and volcanic occurrences. A challenge was to make the book useful and interesting to a wide range of individuals.

The book is organized into an Introduction, ten core chapters, and a chapter entitled Conclusions and speculations. The Introduction conceptualizes the use of the temporal and permanent imprint produced by even moderate sized earthquakes and why past occurrences merit consideration by archaeologists and historians. Chapter 2 describes the earthquake geography of the Americas. Chapter 3 examines the historical role of myths and legends. Chapter 4 describes the interplay between seismology and the search for archaeological features diagnostic of earthquake occurrences. Chapter 5 discusses the occurrence of earthquakes in Mexico and what can be gleaned from the very early written record. Chapter 6 examines the evidence for earthquake occurrences in the Maya Empire. Chapters 7 to 11 offer discussions of earthquake occurrences in other geographical regions of the Americas. Several Appendices summarize observational data relating earthquake magnitude to various dimensions attributed to a specific value of Modified Mercalli intensity. One should note that even though some citations to source materials are found within the text, all relevant sources used are to be found in the chapter-by-chapter Bibliographic Summaries. The References following the Bibliographic Summaries list the full citations for those readers who want to delve further into specific topics.

Special thanks are due to the following organizations who have graciously permitted the reproduction of material in this book: American Geographical Society, American Geophysical Union, Arizona Geological Survey, Arizona Historical Society, Bibliothèque Nationale de France, Carnegie Institution of Washington, Libreria José Porrua Turanzas, Geological Society of America, Peabody Museum Press, Seismological Society of America, and the University of Pennsylvania Museum.

Any book is the product of more than one person. Many individuals have helped me in the search for reference materials and photographs, especially the staff of the Branner Earth Sciences Library at Stanford University. My colleague Amos Nur suggested the development of a book together with leading seminar expeditions to Mexico. Margaret Muir steered me through many technical formatting difficulties. The enthusiasm of my student Bernabe N. C. Garcia, Jr., who spent several months in Mexico and Peru gathering information and photographing many ruins, particularly Palenque, Monte Albán, and Mitla, is gratefully appreciated. I particularly want to acknowledge the assistance and help of my science editor, Susan Francis. Special thanks are due to Mandy Kingsmill for her expertise in editing and improving many aspects of the text. Final thanks are due my wife Linda whose clear-headedness, encouragement, and computer skills brought this effort to fruition.

# 1

## Introduction

Seismology, the investigation of earthquakes and their effects, developed later than most sciences. Instrumental measurements only date back to the late nineteenth century. Earlier historical records of seismic occurrences are pieced together from written descriptions, earthquake catalogs, and the like, and are most useful from the eighteenth century onward. Unfortunately, however, earthquakes usually attracted attention only when man-made constructions were destroyed. Little care, except for describing secondary effects of shaking, was devoted to more careful studies that would shed light on the tectonic processes involved or the reasons that the earthquake occurred where it did. It is not at all surprising that much of the earlier fieldwork done by investigators, such as geologists and archaeologists, may have focused on other matters and as a result significant details pertinent to the earthquake problem were overlooked.

Early studies of the geography of earthquakes relied solely on *macroseismic* data – observations based on descriptions from a single destructive earthquake or on a number of earthquakes felt at a given locality. Mallet and Mallet (1858) compiled a catalog of the world's earthquakes and published a remarkable colored map showing the major earthquake zones and volcanoes of the world. The map, representing a global visualization of seismic geography, is surprisingly accurate when stared at today but it seems to have been ignored in its time. A portion of this map, centered on the Americas, is shown in Figure 1.1. Regions where either the number, or the felt effect, or both, of successive earthquakes are the greatest are shown in the original with a deeper orange-red tint. Volcanoes and fumaroles active in recent geological time are indicated with solid black dots. Areas shown without tint were not presumed to be free from earthquakes but rather were lacking observations. It should also be noted that the color tint is more intense in narrow bands circumscribing the linear alignments of

**Figure 1.1** A portion of the Seismographic Map of the World prepared by Mallet and Mallet in 1858.

volcanic vents. The main features of earthquake locations and their depths of occurrences were fairly well known by the 1930s, culminating in the classical works of Gutenberg and Richter (1941, 1949).

It was the idea of plate tectonics that put earthquake geography in its proper perspective. Most of the energy expended at the surface of the earth is spent as differential movements between plates, and takes place within a few narrow seismic belts. We now know why earthquakes occur where they do and what their main mechanisms of failure are. The beauty of plate tectonics is that it not only explains contemporaneous tectonic activity but it also shows that activity that is going on now must have also been going on for a considerable period of time. On a time scale of hundreds of years the average repeat time of large earthquakes, expressed as jerks in the movement of plates, can be estimated. Even if large earthquakes cluster in time as earthquake storms, the seismic slip rate deduced from the movement of tectonic plates often permits estimation of the average return time for large earthquakes. Can we find corroborating evidence for the past occurrences of such seismic events?

Did indigenous native cultures, Indians of the Pacific northwest, the Aztecs, Mayas, and Incas, to name a few, document the occurrences of natural disasters? Written history, if it can be found, is much less subject to change than oral tradition, provided the original writer was accurate and reliable. However, many occurrences may have only been recorded as myths and legends, and orally transmitted. Stripped of their allegorical cloak, myths of creation and destruction by natural events such as floods, earthquakes, and volcanic eruptions of successive "worlds" are useful because they tell us something of the past and moreover confirm our suspicions that urgent messages were being transmitted into human existence. Myths were used to relay information about past or future occurrences stressing the risks and consequences of natural hazards, and perhaps preparations for interesting times yet to come.

Written records of earthquakes and other natural disasters in the Americas, which occurred prior to the Spanish conquest, are extremely sparse. Overzealous friars destroyed most of the pre-conquest written records, the native Mesoamerican manuscripts referred to as codices. The few pre-conquest Maya and Mexican codices that have been found primarily deal with deities, rituals, the passage of time, world directions, and genealogies. It is not until immediately after the conquest, when native scribes under Spanish sponsorship prepared codices such as the Codex Aubin and Codex Telleriano-Remensis, that specific earthquakes, volcanic eruptions, and the like are mentioned. The earliest occurrence depicted in a codex in pictographic form is an earthquake in 1460 described in the Codex Telleriano-Remensis.

The lack of any historical written tradition is exemplified in the Inca Empire. At least so far as is known there was no system of writing in the Andes in or before the Inca Empire, which reached its zenith about the time of the Castilian conquest in the early sixteenth century. One used the oral tradition for the transmission of knowledge. Even though myths and legends may implicitly contain information about past natural disasters, answers may also have been left in the archaeological record or as colloquially stated "found in the ground." Stated in another way, even if written and oral history can tell us of events and actions, archaeology can often capture the effect of these events and actions.

In studying historical earthquakes one makes use of data extracted from a seismic intensity scale, most commonly the Modified Mercalli Intensity Scale (described in detail in Chapter 4), a subjective numerical scale ranging from 1 to 12 and indicated by roman numerals from I to XII. A number of intensity scales have been used throughout the years and they have been applied to descriptions of a very heterogeneous nature. It is possible, however, to convert these observations to Mercalli intensity values or ranges of values. The basic use of intensity as a descriptive shorthand of earthquake effects collapsed into a single number needs to be kept in mind. However, its concept is useful because isoseismal maps often form the basis for seismic zoning and maps introduced into earthquake design regulations and codes.

Intensity VII is usually taken to be the threshold of significant damage in unreinforced masonry structures such as adobe. The falling of cornices or other architectural ornaments is also typical. Intensity VIII would be assigned to areas showing partially collapsed masonry buildings, the falling of some masonry walls, and the twisting and falling of monuments. Considering that many ancient structures may not initially have been designed with earthquakes in mind, an intensity value of VII could easily have been an VIII. Nevertheless, a seismic intensity value of VII is a useful minimum value which might be inferred from studies of archaeological ruins. Regardless of its shortcomings, intensity values can be roughly correlated to some physical quantity, usually the maximum horizontal ground acceleration that took place during the earthquake. Estimated Modified Mercalli intensity maps based on instrumental ground motion acceleration recordings are now routinely constructed for southern California earthquakes in quasi-real time. It is generally assumed that an acceleration of 0.1 g (1 g = 980 cm/s$^2$), or 10 percent of the Earth's gravitational acceleration, is near the level of ground acceleration where damage to structures of weak construction begins to be noted.

Recently, studies have been made of groups of rocks that have been precariously balanced for thousands of years on many rock outcrops in the western United States. Brune (1996, 1999, 2002) has shown that the positions of precarious and semiprecarious rocks provide a constraint on the level of ground shaking that took place in the historical past. Precarious rocks are defined as those rocks capable of being toppled by peak accelerations ranging from 0.1 to 0.3 g. These levels of peak accelerations would correspond to intensity levels of at least VII. Semiprecarious rocks can be defined in a similar manner but with peak accelerations of 0.3 to 0.5 g required for toppling. One main use of the studies of precarious and semiprecarious rocks is that they provide data for evaluating the level of ground shaking to be expected for earthquakes possessing long recurrence or repeat times. Debate exists, however, in estimating the length of time that rocks existed in their precarious state and why precarious rocks are not always found where they might be expected.

Empirical correlations between seismic intensity and earthquake magnitude can be made (Appendices A and B). Magnitude can be visualized as the power of a seismic transmitter, the size of the earthquake, whereas intensity should be imagined as the strength of the signal received at a particular observation point. Intensity is a subjective number that can be loosely correlated with the observationally determined magnitude assigned to an earthquake. For example, a magnitude 8 earthquake might generate maximum seismic intensities of XI or greater and be felt at distances of 800 km (~500 miles). Magnitude 6 earthquakes can be felt at distances of several hundred kilometers and produce damaging intensities ranging from VII to VIII, the value where common damage to structures is noted.

The conventional earthquake magnitude scale $M$, is useful to describe the size of an earthquake but it has a shortcoming in that the scale saturates for very large earthquakes. The reason that saturation occurs is that very large earthquakes release energy at low frequencies that are not recorded by conventional seismographs. Seismologists now use a moment-magnitude, $M_w$, scale, whose value is related to the fault area which has ruptured, the amount of fault slip or offset across the fault, and the strength of the rocks involved. The difference between $M$ and $M_w$ is only critical for very large seismic events.

In delving further into the past history of earthquake occurrences one can also resort to *paleoseismology* or studies that examine the geological signature left by past earthquakes. Of particular interest are events that can be dated and that occurred in the Holocene, a time span encompassing about 10 000 years before the present. When a large earthquake has taken place it can produce geological features such as fault scarps, offset stream channels, regions of uplift, downdrop or tilting, landslides, and soil liquefaction. Depending upon the environment these features are often preserved in the geological and/or stratigraphic record and can be recognized hundreds, and at times thousands, of years later.

Even though information about past earthquake occurrences can often be "found in the ground" their acceptance as a major cause of changes in archaeological horizons has not been universally accepted. Typically such changes are explained in cultural terms such as armies or conquests, missionizing movements and conversions, social and political decay, or agricultural disintegration. Tectonic, seismic, and volcanic events often produce natural disasters affecting large geographic areas. These disasters and other marine and meteorological events may have left clues in the archaeological record. The blending of archaeology and geological observations to decipher past occurrences can be illustrated using an example from Death Valley, California. A Holocene fault scarp at the foot of the Black Mountains, on the east side of Death Valley, was a site occupied no earlier than AD 1 by the Death Valley III Indians. Indian mesquite pits with artifacts are found on both sides of the fault and on the eroded escarpment itself (point D on Figure 1.2). Desert varnish, an iron and manganese oxide, is a deposit of considerable antiquity but it has not been deposited during the past 2000 years and is not found on the Death Valley III artifacts. Geological faulting, evidenced by the Holocene discontinuous escarpment, although now 90 percent destroyed and buried under younger alluvial fan deposits, and eastward tilting

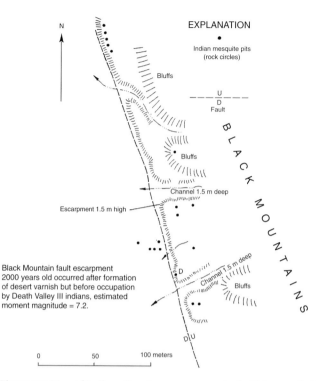

**Figure 1.2** Map of Indian sites along escarpment of a Holocene fault at the western foot of the Black Mountains, Death Valley. Fault and escarpment are older than the Indian mesquite storage pit at D, which was built on the colluvium overlapping the scarp (Hunt and Mabey, 1966).

of the neighboring salt pan, is known to have taken place about 2000 years ago, after the ubiquitous desert varnish was formed but definitely earlier than the local Death Valley III occupation. It is straightforward from the length and amount of fault offset to deduce that the moment magnitude of the earthquake that produced the vertical displacement along the front of the Black Mountains had to be at least 7.2.

The acceptance of earthquake damage at archaeological sites has had a checkered history with advocates both for and against. Los Muertos, located about 16 km south of the present town of Tempe, Arizona, was the largest Salado Indian community in southern Arizona in pre-Spanish times. The site was permanently abandoned sometime during AD 1400–1450. During excavations by the Hemenway southwestern expedition of 1887–1888, a skeleton of a man caught beneath a collapsed wall was unearthed and cited as evidence of destruction by an earthquake. At the time, F. H. Cushing, the leader of the expedition, was firmly convinced that earthquakes were responsible for the demise of the community:

Why were some of these Southwestern systems of cities abandoned so long ago, while others remained occupied within comparatively recent times, and still others until even the present day, as is the case with the Zuñi descendants of these primal ruin-builders? The answer to this question is . . . abandonment mainly through earthquake disturbances. That such disturbances were the cause of the abandonment of at least the lower Gila and Salado cities, seems indisputable, to my mind, after a careful examination which I was enabled to make of their condition and distribution of the remains they contained, and especially of the occurrences there of earthquake sacrifices, kindred to, though much more extensive than those made in modern Zuñi on occasion of even slight earth-tremors or landslides . . . This series of facts becomes still more potent in explaining the causes which led to the abandonment of the Lower Gila Salado cities, when in them we find – as tradition states of the ancestral towns of the Zuñis – that there were abandoned within their walls all that was best; and when we find long rows of houses, in certain directions, tumbled down in true earthquake fashion, the roofs burned by the hearth-fires that were burning even when they fell, skeletons crushed under them, and finally, more significant than all, actually, at least in some of these cities, Earthquake Ceremonial Appliances – identical with those of Zuñi – as was the case in one of the sacred lodge-rooms we chanced to open at Los Muertos. (Cushing, 1890).

Yet the explanation offered in a much later summary report (Haury, 1945) was not supportive of earthquakes as the cause of the exodus of the Salado: "But the finding of tumbled down houses and burnt roofs – the evidence adduced for earthquake destruction – is of such common occurrence in all parts of the Southwest and represents so many horizons that the earthquake theory seems quite inadequate."

Earthquakes were felt by early indigenous people of the Americas and they formed a part of the world of early Native American myths and legends. The Maricopa Indians, a Yuman tribe of southern Arizona, lived in a region near the confluence of the Gila and Salt Rivers, approximately 22 km west of Phoenix and 35 km west of Tempe, Arizona. *Cipas* was one of the twin culture gods of creation in the lore of the Maricopa. When he died he went under the earth where he still lies. Whenever he is tired he yawns and stretches or turns over causing an earthquake (*mathenk*, "earth shakes"). Earthquake occurrences may or may not have been viewed as particularly ominous but were not completely overlooked. The Maricopa community used wooden calendar sticks marked with notches about 1 to 1.5 cm apart signifying years (typically beginning early February with the budding of trees). Certain historical events such as battles or pestilence were noted with additional cryptic markings. A calendar stick examined by Spier (1933, p. 141) covering an interval of time from 1833 to 1930 documented the occurrence of the 1887 Sonora, Mexico, earthquake discussed in Chapter 10, which was felt over much of the American southwest and northern Mexico.

Another example of an indigenous Indian culture, which evolved, flourished, and declined under a very dynamic landscape is the Maya of Middle America. Their geographic location with respect to the sliding Pacific and Caribbean plates, and their proximity to the volcanic spine of Central America undoubtedly made them vulnerable to natural events of local and regional extent. Evidence of collapsed vaults (even those that were buttressed) and stelae that are toppled in preferential directions or systematically leaning have not been generally accepted as the consequences of earthquake occurrences. Instead the found conditions have been solely attributed to vegetation and fallen trees. Archaeological expeditions have been intent on the righting, repair, and restoration of monuments and clearing of the site to pristine conditions, without critically examining the possibility of earthquake damage. The fact that earthquakes *may* have caused damage at many Maya sites has not gained currency or credibility and is usually dismissed as an unimportant contribution to abandonment or rebuilding:

> Concerning the first or earthquake hypothesis, it appears to be the most improbable of all. It rests primarily upon the present ruined condition of the Old Empire cities, the fallen temples and palaces, and the overthrown and shattered monuments, and the prevalence of severe earthquakes in adjacent areas . . . since the vegetation now covering the sites of the Maya cities is alone sufficient to account for their destruction, the writer believes the seismic hypothesis may be rejected. (Morley, 1920, pp. 442–443)

With a lack of consensus, however, on the most important factor for the demise of the Maya, it has been proposed that it was a gradual phenomenon, triggered by a series of interrelated factors, that led to its downfall. Different interpretations of these factors involve civil wars, invasions, plagues, hurricanes, earthquakes, overpopulation, agricultural exhaustion, or a breakdown of trade networks. Even if one does not agree that earthquakes caused the collapse of cultures, one should not singularly exclude the possibility that they are a contributing factor. Clues are often metaphorical. In the book of *Chilam Balam of Tizimin*, a historical text of the Yucatecan Maya dominated by cyclical repetition of the katun, a period of 7200 days, reference is made to an earthquake in the year bearer 4 Kan (AD 1597):

> On this fourth Kan,
>     The day
> Of the movement of heaven,
>     The movement of earth,
> Knocking together the stubborn sun
> priests,
>     Knocking together the lands in the
>     country,
> The pruning of the *katun*
>     When it is seated. (Edmonson, 1982)

(The translator did not believe that it was an earthquake being discussed, perceiving the sense to be figurative since Yucatán was visualized outside the earthquake zone.) The *Maní* variant of the *Book of Chilam Balam* (Craine and Reindorp, 1979) translates this phraseology differently, "On 4 Kan of the *katun* 5 Ahau . . . Trees were broken, rocks were split, the earth burned, and frogs croaked in the wells at midday . . . Another language would come, and the sky and the Earth would tremble in the Petén."

Infrequent earthquakes have been *felt* in the Yucatán peninsula. On November 15, 1908, two earthquakes occurring late in the evening were felt across the Yucatán peninsula. Earthquake shaking with a duration of 15 to 20 s was experienced from Chetumal and Vigia Chico on the Caribbean coast, as far east as Hecelchakán and Iturbide in the state of Campeche, and in the communities of Merida, Ticul, Tekax, and Izamal in the state of Yucatán (Figure 1.3). The earthquake was of sufficient intensity to cause great alarm and panic and can justifiably be assigned a minimum Modified Mercalli intensity value of VI. The minimal area subjected to this level of intensity is $50\,600$ km$^2$ suggesting at least an $M = 6.5$ earthquake. Occurrences of earlier comparable earthquakes in the historical past cannot be summarily dismissed.

Two fault zones are known, on the basis of topographic alignments, to be present in the Yucatán, the Ticul, and Rio Hondo fault zones. No evidence for past earthquake occurrences has yet been found on these faults but they have not yet been monitored for current microearthquake activity and cannot be excluded as a past earthquake source. A seismic station at Merida, in operation since 1911, is not designed for local earthquake monitoring.

A number of criteria are useful for identifying earthquake effects from archaeological observations. Isolated instances of damage have to be assessed within the overall framework of an archaeological and seismotectonic setting. For example,

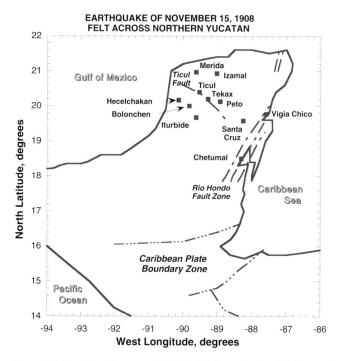

**Figure 1.3** Location of communities in the Yucatán that felt the earthquakes of November 15, 1908.

collapsed walls could be the end result of weakening from fires, the result of invading conquerors pushing down the walls, the consequences of neglect and vegetation effects, or simply the result of poor construction. Skeletons found under walls provide gruesome evidence for collapse from earthquake shaking. Pottery and other artifacts, that could have been looted, found on the floor of structures are further indicators of sudden collapse. Earthquakes are usually not accepted as the cause of much of the damage noted at many Mesoamerican ruins. However, many of the damage features observed strongly suggest that they were indeed produced by earthquake shaking. Topographical, climatic, and anti-seismic considerations must have intervened in the varied physiognomy of the structures found at many Mesoamerican centers.

Archaeological evidence points to the sudden, possibly catastrophic, abandonment of some Maya sites. We should not summarily dismiss the possibility that the observed breakage and toppling of Maya stelae could have been the result of natural causes *during* and *after* final abandonment. Some stelae have also been found reset upside down or only re-erected as top fragments, which implies an earlier breakage. The intentional ancient throwing down of a stela for hostile reasons by undermining its foundation is of course not impossible

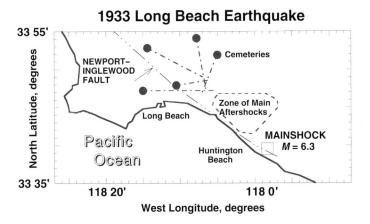

**Figure 1.4** Map showing the epicenter and aftershock zone of the $M = 6.3$ Long Beach earthquake of March 10, 1933. The solid circles north of Long Beach are locations of cemeteries. Assuming transverse motion to have produced the fall of the tombstones the dashed lines drawn perpendicular to the average fall directions converge near the northwestern end of the rupture zone.

but the reasons for any resetting and the length of time between breakage and resetting are continuing and puzzling problems in Maya archaeology.

In the spirit of being neither pig-headed nor disputatious one has also to accept as an observational fact that many ancient monuments in the Maya empire were initially found not to be level. Does one conclude that the ancients congregated for the ritual positioning of a crooked monolithic stela or do we examine the possibility that earthquakes could have played roles in their tilting and toppling? Some coherency in the sense of prevailing shaking in one direction is not surprising considering the wavelengths of the seismic waves involved. Wavelengths of the shear or transverse seismic waves, which produce the heavy shaking, can be of the order of tens of kilometers. Horizontal accelerations produced by earthquake shaking could easily have contributed to the destruction of buildings, walls, and other structures at ancient Mesoamerican sites, which then contributed to subsequent political and social collapse.

The concept of monuments falling in a prevalent direction can be illustrated by examining the effects of the Long Beach earthquake of 1933. This earthquake, with its epicenter near Huntington Beach, California, ruptured along the Newport–Inglewood fault zone in a northwesterly direction towards Long Beach, where most of the aftershocks were clustered (Figure 1.4). After the earthquake(s), graveyard tombstones were noted to have fallen in preferential directions. It can be seen that lines drawn perpendicular to the majority of the preferred fall directions converge near the northwest end of the rupture zone supporting the

**Figure 1.5** Damage to the Inglewood Hotel from the earthquake of June 21, 1920. The brick front wall was not properly tied to the frame of the building and fell outward exposing the rooms on the second floor and crushing a small automobile in the right foreground (Taber, 1920).

idea that gravestone toppling was the result of the large-amplitude transverse or shear waves generated by the earthquake and/or its aftershocks.

Even a moderate-sized earthquake can cause damage to poorly constructed buildings and the toppling and movement of monuments in graveyards near an epicentral area. On June 21, 1920, an earthquake occurred in Inglewood, California, about 16 km southwest of downtown Los Angeles. Modified Mercalli Intensities of VII to VIII were assigned to the area of maximum damage. The typical damage to buildings was the result of poor construction where thin brick walls built as fronts to wooden-framed structures fell outward because they were not properly tied to the framing (Figure 1.5). Of more interest was the observation that monuments were displaced and toppled in a nearby cemetery (Figure 1.6). Analyses of the size of the areas subjected to intensities of VII to VIII suggest a magnitude in the range of 4.7 to 5.2. This earthquake was subsequently assigned a magnitude of 4.9 showing that even a minor earthquake can topple and tilt monuments at close distances to a causative fault. It is the level of horizontal acceleration that is important. Ground accelerations and velocities can be unusually high in the vicinity of earthquake faults when a propagating

Figure 1.6 Overturned and displaced monuments in the Inglewood Cemetery as a result of the shaking from the Inglewood earthquake of June 21, 1920 (Taber, 1920).

rupture is involved because of a directivity or focusing effect. For example, the relatively modest-sized $M = 6.6$ Imperial Valley, California, earthquake of 1979 produced transverse ground accelerations ranging from about 0.4 to 0.8 $g$ and vertical accelerations greater than 1 $g$.

In the chapters to follow we will span a number of geographic areas and examine earthquake occurrences, both past and present. In *sensu stricto* not every region of the Americas is covered in detail. Rather the intent has been to include regions where important earthquakes have occurred, speculate on their impact and historical role, and examine the record left by the transient event. There remains much to study as the combination of earthquakes and the archaeological record is an evolving discipline. Let us begin with an overview of the geography of the earthquakes of the Americas.

2

# Earthquake mosaic of the Americas

## 2.1  Some preliminaries

The earthquake activity of the Americas is readily understood when it is viewed as a segment of the circum-Pacific belt; 80 percent of the world's earthquakes take place (Figure 2.1) in this zone. In discussing the geography of seismic activity we need to be aware that earthquakes can occur at various depths of focus or points of origin within the earth. A shallow-focus earthquake has a focal depth or hypocentral depth ranging from 0 to 70 km. Focal depths of intermediate-focus earthquakes range from 70 to 300 km and deep-focus earthquakes range from about 300 to 800 km. There are no earthquakes with focal depths exceeding 800 km.

Maps of seismic activity often show earthquake locations discriminated on the basis of their size or magnitude, $M$. Gutenberg and Richter (1941) placed earthquakes into various magnitude classes. Class **a** earthquakes were assigned a magnitude range of 7.8 to 8.5 or greater, class **b** $M = 7$ to 7.7, class **c** $M = 6$ to 7, and class **d** $M = 5.3$ to 6. Earthquakes with a magnitude greater than 7.5 are often called *great* earthquakes, those with magnitudes from 6.5 to 7.5 *major*, and those ranging from 5.5 to 6.5 *large* earthquakes. A magnitude 5.5 earthquake is considered to be the minimum size of event to produce a threshold of significant damage. Earthquake magnitude is an instrumentally determined parameter and can only be estimated by damage assessment for historical events. There are many magnitude scales in existence. When discussing historical events, $M$ refers to the Richter magnitude. Because these values can be somewhat misleading for very large earthquakes seismologists also use a surface wave magnitude, $M_s$, or a moment magnitude, $M_w$, to help achieve a uniform comparison of earthquake size. These scales can be related to dimensions of fault rupture through

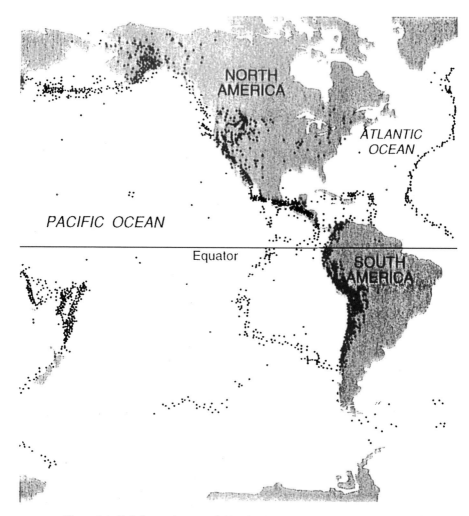

**Figure 2.1** Global mosaic map of the seismic activity of the western hemisphere.

a parameter known as the seismic moment, $M_0$. Useful empirical relations connecting the seismic moment, in dyne cm, with earthquake magnitude are:

$$\log M_0 = 1.5M_s + 16.1$$
$$M_w = \frac{2}{3}\log M_0 - 10.73$$

Before instrumental recordings began near the beginning of the twentieth century, it was realized that the effects of a past earthquake could be classified according to the apparent intensity of its tremors noted at different locations. Intensities at various points were thus quantified according to a closed-end

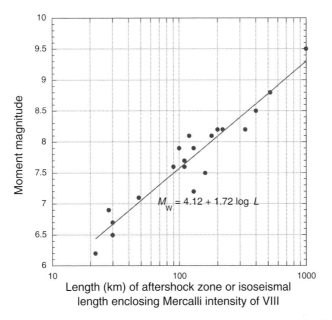

Figure 2.2 $M_w$ versus log $L$ (rupture length or length enclosed by Mercalli intensity VIII in km) for Californian, Central American, and South American earthquakes.

numerical scale ranging from I to X (Rossi–Forel) or I to XII (Modified Mercalli) and an *isoseismal* map was prepared (see Chapter 4). From such felt observations, examination of earthquake ruins and building damage, written descriptions and, at times, preserved geological features, the size of the historical earthquake can be adequately determined for magnitude assessment and comparison purposes. Estimates of the size of historical events can also be made from isoseismal contour areas or lengths enclosed by a specific intensity level that have been calibrated against twentieth-century earthquakes of known magnitude. The maximum length of the isoseismal contour or area enclosing Modified Mercalli intensity VIII, for which adobe structures in the Americas suffered severe-to-complete destruction, correlates well with the length, $L$, of the earthquake rupture zone defined by the aftershock activity for earthquakes of the Americas of known size. Figure 2.2 shows such a correlation.

In our broad overview of the seismicity of the Americas it is convenient to discuss earthquake occurrences that take place in broad, geographic areas. The regional areas are not meant to be pedantic or arbitrary but simply to offer a convenient framework for discussion. These regions are: (1) the northwest Pacific zone, including California; (2) the intra-continental area of North America;

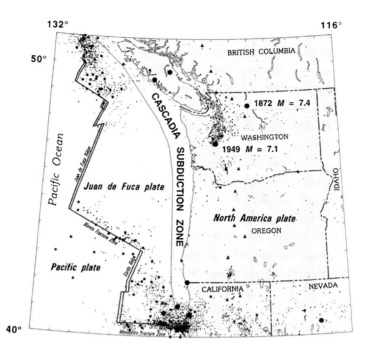

**Figure 2.3** Epicenter map (solid circles) for the Pacific northwest for the period 1833–1993. Triangles are volcanoes. Location of 1872 and 1949 events labeled. Observe the absence of current seismicity along the Cascadia subduction zone (annotated from Rogers *et al.*, 1996).

(3) the western coastal zone of Mexico and Central America; (4) the Caribbean zone; and (5) the coastal South American and Andean zone.

## 2.2    Northwest coastal Pacific zone and California

The seismic activity of the northwest coastal Pacific zone, particularly in the vicinity of Washington and Oregon, is relatively low when compared to other portions of the circum-Pacific belt. Seismicity is still appreciably higher than that of eastern North America, and is often accented by large-magnitude shocks, such as the 1949 $M = 7.1$ Puget Sound event, and the 1872 $M = 7.4$ event in north central Washington, which appear to recur at comparatively long time intervals. With the exception of the Puget Sound–Seattle area, most of the seismic activity in the zone from British Columbia southward to the Mendocino fracture zone near northern California takes place offshore (Figure 2.3).

Even with the presence of the Cascadia subduction zone, where the Juan de Fuca plate descends beneath the North American plate, all of the seismic activity is shallow focus. Along the Cascadia subduction zone itself there has been a

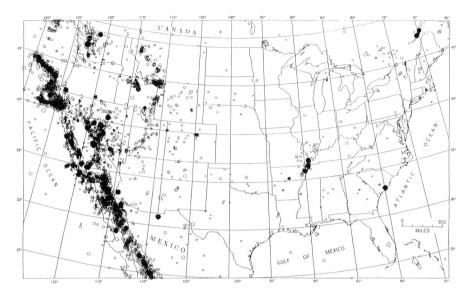

Figure 2.4 Seismic activity of the United States.

noticeable absence of shallow coastal seismic activity over the past 300 years. The subduction zone is not completely benign. Aseismic slip may be taking place along the plate boundary with the infrequent occurrence of large-magnitude events on the time scale of hundreds of years. Paleoearthquake evidence, such as the presence of offshore turbidites, buried soils, and dead forests, which were produced by sudden coastal subsidence and liquefaction of sandy soils due to ground shaking, suggest the occurrence of a past great offshore earthquake(s) that presumably originated in the subducting plate. Such an event is believed to have taken place in January, 1700, corroborated by the presence of sand layers, deposited by a tsunami covering dated Indian fire hearths. This event is also alluded to in North American Indian legends.

From a latitude of about 40° N southward to a latitude of 25° N the west coast of North America is dominated by the band of earthquake activity produced by the San Andreas fault system (Figure 2.4). This zone of seismicity passes south-easterly paralleling the coast of California to just north of Los Angeles where it bends to the east between latitudes of 34° and 35° N for several hundred kilometers before turning southeasterly again through the Imperial Valley and into the Gulf of California. The written record of California earthquakes only extends back several hundred years or so. Two great earthquakes have occurred on the San Andreas Fault within this time interval, in 1857 and 1906. Major earthquakes took place in 1812 and 1838. An $M = 6.9$ earthquake, known as the Loma Prieta event, took place on October 17, 1989, on a somewhat remote segment of the San Andreas Fault in northern California. There were 62 deaths but extensive damage to property and infrastructure in the San Francisco Bay area.


## 2.3    North American intra-continental areas

Earthquakes in western North America, excluding the coastal Pacific zone, take place in a broad mountain region passing northward into Idaho and Montana from the California–Nevada border eastward as far as central New Mexico and Colorado (Figure 2.4). One event, the Sonora earthquake of 1887, with an epicenter just south of the joint border between Arizona and New Mexico, in a sparsely populated desert expanse, is often overlooked in terms of the geological changes that accompanied it as one of the great earthquakes of the Americas. Other earthquakes of interest in the western mountain region, discussed in Chapter 10, were the $M = 7.75$ Pleasant Valley, Nevada, earthquake of October 2, 1915; the $M = 7.1$ Hebgen Lake, Montana, event of August 17, 1959; and the $M = 7.1$ Dixie Valley earthquake of December 16, 1954.

East of the Mississippi River major seismic events have taken place in three principal regions. These regions are the St. Lawrence River valley in the northeast, the South Carolina area of the Atlantic seaboard, and the Mississippi River area of southeastern Missouri, centered at a latitude of about 37° N. Two major earthquakes of historical importance that occurred in the St. Lawrence region were the St. Maurice earthquake of February 5, 1663 and the $M = 7$ St. Lawrence earthquake of March 1, 1925. The earlier earthquake, with its presumed epicenter about half-way between Montreal and Quebec, on a tributary of the St. Lawrence was described by the French Jesuits of the era as a tremble-terre (a quaint colloquialism of tremblement de terre). The 1925 event, sometimes referred to as the La Malbaie, Quebec, earthquake, was felt as far south as Virginia and over an area of 360 000 km². However, all significant damage was confined to a narrow belt covering both sides of the St. Lawrence River.

Charleston, South Carolina, not considered as the locus of frequent, damaging earthquakes, was shaken by a violent earthquake on September 1, 1886. The number of fatalities was small but it was perceptible over at least an area of 1 million km². The magnitude of this event has been estimated to be comparable to the New Madrid earthquakes of 1811–12 in southeastern Missouri, but the latter events produced much more severe damage in the epicentral area. The epicentral area of the New Madrid earthquakes encompassed a band of land of 11 000 to 18 000 km² bordering the Mississippi River, including parts of western Tennessee and northeast Arkansas. One of the most conspicuous effects of this earthquake was the formation of a vast zone of subsidence with the creation of

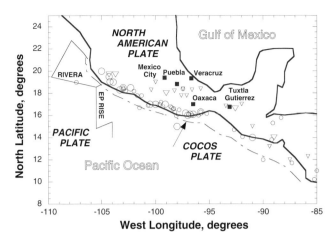

**Figure 2.5** Seismic activity of Mexico and a portion of Central America from 1900 to 2000. Circles are shallow-focus events and open inverted triangles are intermediate-focus events. Larger-sized symbols are for events of $M \geq 7.8$. Smaller symbols are for events in the magnitude range 7 to 7.7.

new lakes, swamps, and bayous. The seismic events of central and eastern North America will be discussed in greater detail in Chapter 11.

## 2.4    Western coastal Pacific zone of Mexico and Central America

The Pacific coastal region of central Mexico and Central America, from Jalisco at a latitude of 20° N southward to a latitude of about 10° N, is one of the more active segments of the circum-Pacific belt, particularly for shallow and intermediate focal-depth earthquakes (Figure 2.5). Volcano locations in Mexico and Central America parallel the western coast emphasizing the spatial corre- lation between earthquakes and volcanic activity in the familiar "Pacific ring of fire." Shallow-focus activity roughly parallels the Pacific coast turning more southward offshore of Panama. Inward of the west coast and passing southeast- erly through Central America, the main zone of seismic activity is primarily intermediate focal-depth (~100 km) events. A conspicuous west-to-east trend of intermediate-focus events is also present at a latitude of about 19° N. This zone of intermediate focal-depth events follows a trend of very active volcanic activity across Mexico. At a latitude of about 15° N near Tehuantepec, the seismic belt branches east-northeasterly continuing as the Caribbean seismic loop. Earth- quake occurrences along the Mexican and Central American coasts are dictated by the north-northeasterly convergence and subduction of the Cocos plate be- neath the North American plate. Damaging shallow-focus and intermediate-focus earthquakes are aligned along the strike and dip of the subducting plate.

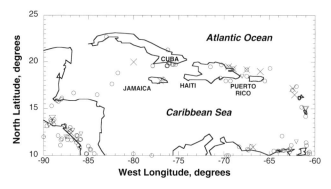

**Figure 2.6** Seismicity of the Caribbean loop from 1690 to 2001. Larger-sized open inverted triangles are intermediate-focus events of $M \geq 7.8$; smaller inverted triangles are events in the magnitude range 7 to 7.7. X marks shallow-focus events of $M \geq 7.8$; open circles are events in the range 6.0 to 7.7.

Mexico has a long history of damaging earthquakes. Mexico's seismic history can be illustrated with a simple comparison to California. In the twentieth century, California had experienced 5 earthquakes of $M = 7$ or greater; Mexico had experienced 42. This fact should be borne in mind when one is examining the seismic history of Mexico and its probable effect on the indigenous population. A very destructive earthquake struck Mexico City on September 19, 1985. This $M = 7.9$ event had its epicenter off the Mexican coast in the Michoacán area some 350 km from Mexico City; 9500 people were killed with many more injured and made homeless. Property damage totaled 4 billion dollars. The epicenter was distant from the capital but caused considerable damage because parts of modern-day Mexico City are built on an ancient lake bed, which was drained by the Spanish after the conquest of the Aztecs. The old lake bed contains soft deposits of clay and sand, and seismic waves of certain frequencies are preferentially amplified. Buildings and structures with similar natural frequencies of vibration related to their heights can be driven to large oscillations because of resonance.

## 2.5    The Caribbean loop

Most of the earthquakes of the Caribbean are confined to a belt that can be traced from Central America to the Greater and Lesser Antilles and thence to South America (Figure 2.6). Seismic activity is highest near the island of Hispaniola and in northeastern Venezuela. Most of the activity is produced by shallow-focus events. However, a zone of intermediate focal-depth earthquakes can be followed along the length of the Lesser Antilles. Intermediate-focus earthquakes are also found under the eastern end of Hispaniola.

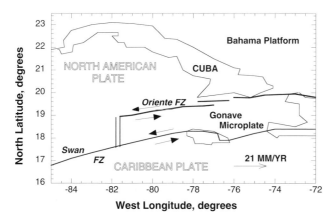

**Figure 2.7** Kinematic setting of the Caribbean plate boundary near Cuba.

On the western end, the zone of activity extends from the Yucatán peninsula and Honduras east-northeasterly along a lineament known as the Bartlett Deep or the Swan fracture zone. This feature runs towards Cuba and Jamaica. Along this seismic plate boundary the sense of motion is left-lateral strike-slip. South of Cuba motion is accommodated along two fracture zones. The main locus of seismic activity is associated with the Oriente fault zone, which passes eastward through the northern part of Haiti and the Dominican Republic. A zone of less frequent seismic activity runs eastward through Jamaica (Figure 2.7).

The Lesser Antilles have both shallow and intermediate focal-depth events and also contain many active and destructive volcanoes. Historically the record of seismic activity in the Caribbean loop extends over the past four centuries. Unfortunately the small number of locally destructive shocks gives a false historical impression that the Caribbean region is a highly active zone of seismicity when compared to other parts of the circum-Pacific seismic belt.

Seismic and volcanic activity of the Caribbean region result from the convergence of the Caribbean plate and a segment of the Atlantic seafloor that forms a part of the South American plate (Figure 2.8). Plate convergence started during the Jurassic period about 175 million years ago with the opening of the Atlantic Ocean and the adding of new crustal material along the Mid-Atlantic Ridge. In its early stages of convergence the direction of movement of the Caribbean plate was east-northeasterly as the Atlantic seafloor was pushed westward. As a result, the thicker Caribbean plate was forced over the thinner Atlantic oceanic plate and eventually subducted along the arc of the Lesser Antilles. The rate of convergence, however, has not been uniform throughout geological time and the volcanic eruptive centers have moved from east to west along the subducting arc. Although the broad interaction of the Caribbean and North American plates is known, details of the rate of movement along individual segments of the plate

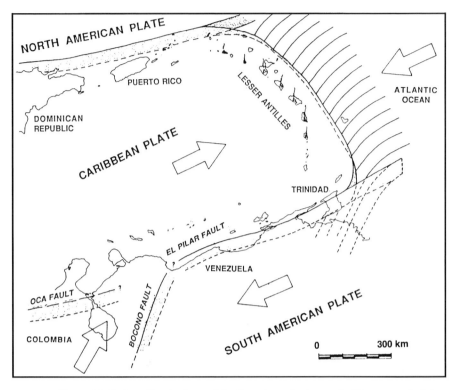

**Figure 2.8** Cartoon illustrating subduction along the Lesser Antilles arc and plate boundary motions between the Caribbean and North and South American plates.

boundary zone are not. The overall rate of eastward movement of the Caribbean plate relative to the North American plate is about 21 mm per year.

## 2.6    South American and Andean zone

The west coast of South America is repeatedly subjected to large-magnitude shocks. Their number and frequency of occurrence are notable. From AD 1900 to 2000, 65 earthquakes of magnitude 7.8 or greater took place in the South American coastal and Andean regions within a rectangular band ranging from 5° N to 40° S latitude. Of this total number, 28 were shallow-focus, emphasizing the continued vulnerability of the coastal area to the effects of large and potentially damaging earthquakes. The earthquake belt parallels the Pacific coast in Chile and follows a northward trend near the edge of the continent along the coastal regions of Peru, Ecuador, Colombia, and into the Central American countries. Earthquake activity in this region defines one of the longest, most continuous seismic zones in the world. The Pacific coast of South America,

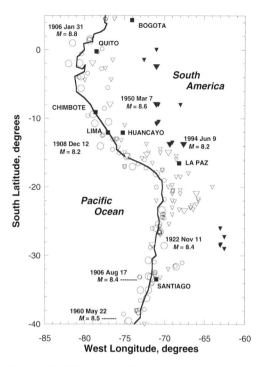

**Figure 2.9** Seismicity of the South American coastal and Andean regions from 1900 to 2000. Circles are shallow-focus events, open inverted triangles are intermediate-focus events, and inverted filled triangles are deep-focus events. Larger-sized symbols are for events of $M \geq 7.8$. Smaller symbols represent events in the magnitude range 7 to 7.7.

together with the Andean region, has shallow and intermediate focal-depth seismic events and is the only location in the Americas that exhibits deep-focus earthquake activity. Deep-focus earthquakes follow a band paralleling the Andean mountain chain. Focal depths systematically increase from the offshore region eastward beneath the Andes reflecting the subduction of the thinner Pacific Ocean lithospheric plate beneath the South American continental plate. Shallow-focus activity is confined close to the coast or offshore. The region has considerable intermediate-focus activity, particularly concentrated at the latitude band of 20° to 23° S (Figure 2.9).

The year of 1906 was notable for the occurrence of two great earthquakes that took place in South America. The first event was on January 31 at a latitude of about 1.5° N and the second was on August 16 at a latitude of approximately 33° S. The January 31, 1906, Ecuador–Columbia event was one of the largest-magnitude shocks ever to take place in the South American zone. On August 16, 1906, a very large earthquake struck near Valparaíso, Chile producing about

3800 fatalities. This earthquake was the subject of a unique compendium, for its time, of seismograms recorded at distant seismographic stations (Rudolph and Tams, 1907).

The great earthquake of May 22, 1960, off the southern coast of Chile paralyzed the country and was a major disaster producing 5700 fatalities. Scientific reconstructions of the size of this event indicate a fault rupture area of 160 000 km² and a fault offset of 21 m! Even though this event was given a conventional Richter magnitude value of 8.5, its moment magnitude was 9.5 to 9.6, making it the largest seismic event worldwide yet instrumentally recorded.

3

## Myths and legends

### 3.1 The historical role

Early cultures developed many beliefs, and it is necessary to understand why these beliefs and explanations about natural phenomena were woven into the fabric of life. History is important, particularly where human nature and behavior are concerned. If this were not true then life would solely be the response to a sequence of genetic impulses. We who choose to trace our origin back to a monkey or believe in, and search for, the Garden of Eden, do not revere a country by chance of birth to the extent that native Indians revere the land where their ancestors originated. This reverence cannot be underestimated. Every hill, body of water, river, or mountain has a story connected with it; an account of its origin, which expresses a view of the nature of the world, often at odds with the perspective of a deterministic science. These stories may be *myths* or *legends*. Myths are traditional stories that focus on the deeds of gods or heroes to explain natural phenomena. Legends, on the other hand, are based on historical events that have been described, and handed down, usually by word of mouth, from generation to generation. Myths and legends represent a means to communicate with the past, often combining outer- and inner-world experiences. The interpretation of myths and legends, as traditional factual accounts of historical personages and events, that can be precisely dated, is usually difficult.

Stories about earthquakes and other natural events such as volcanic eruptions and floods, often discuss choices that are made in the light of uncertainties, namely *hazards*. Hazards, threaded into the texture of myths and legends, form the basis for a language to state simply that as far as human history has unfolded, the world every so often is symbolically destroyed, and a new one created. More specifically, there are many Native American myths that give reasons why the ground trembles. These stories usually involve some creature or divinity that resides in or under the ground, or creatures (usually animistic) that support the earth and shake it for various reasons such as anger and displeasure, or even romantic reasons (one could take this explanation further and state that this is the source of Love waves, a type of propagating seismic wave).

Earthquakes were known to the early American Indians of the east and are a fabric of the Iroquois origin myth of the Turtle, the Earth Bearer, who is restless from time to time, and when he stirs there are earthquakes (Snow, 1994). Early records of earthquakes were often apocryphal. A letter written in 1638 by Roger Williams (1863) to John Winthrop describes a Narragansett story about the New England earthquake of 1638, which was a new experience for the pilgrims: "All these parts felt it [the earthquake] . . . the elldere informe me . . . they allways observed either plague or pox . . . 3, 4 or 5 years after the Earthquake, (or *Naunaumemoauke,* as they speake)." At the time both the colonists and the Indians saw bad omens in the connection between earthquakes and epidemics.

Fear of earthquakes was also a convenient tool for the proselytizing of souls. Father Gonzalo de Tapia, a Jesuit, spent several years amongst the Indians of the Sinaloa nation in Mexico, in particular the Zuague tribe, which resided near Mochicahui on the Rio Fuerte. Mochicahui is located near Los Mochis on the Gulf of California terminus of the famous Chihuahua-to-Pacific railroad. An earthquake that took place in 1593 (specific date not given) was described as splitting open a hill initiating a flow of water. The Zuagues sought to appease the evil spirits by throwing gifts and offerings of esteemed value into the flowing stream of water. They also brought gifts to Father Tapia asking for his interces-sion to the Christian God. The dedicated father used the opportunity to urge them to be baptized (Robertson, 1968, p. 29).

References to earthquakes are found in the ethnohistorical and ethnographic legends of the native peoples who occupied the Washington–Oregon coast par-alleling the Cascadia subduction zone (Figure 3.1). These accounts include de-scriptions of low-lying coastal settlements that were subjected to earthquakes and submerged either by inundation from a local tsunami or tidal flooding. Dated Native-American buried fire hearths found in soils and in river estuaries (Figure 3.2) are compelling evidence for sudden coastal subsidence induced by a great earthquake about 300 years ago along the Cascadia subduction zone. The possible recording of this particular event and earlier known alluvial events is suggested in several northwest Pacific Indian legends.

The Makah Indians who live at Neah Bay, the northwest corner of the territory next to the Pacific Ocean, speak of a very high tide:

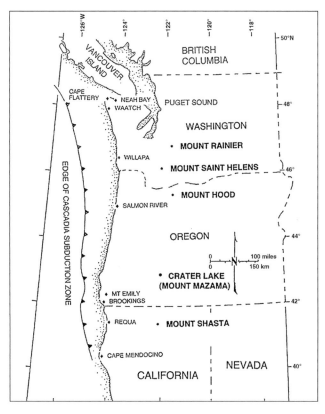

**Figure 3.1** Schematic diagram of the Cascadia subduction zone showing the location of places mentioned in Indian legends. The locations of the main volcanoes of the Cascades are shown.

A long time ago but not at a very remote period the waters of the Pacific flowed through what is now a swamp and prairie between Waatch village and Neah Bay making an island of Cape Flattery. The water suddenly receded making Neah Bay perfectly dry. It was four days reaching its lowest ebb, and then rose again without any waves or breakers till it had submerged the Cape. (Eels, 1878, p. 71)

At Shoalwater (Willapa, about 40 km north of where the Columbia River meets the coast) where evidence of elevation and depression of the land, apparently at no very ancient date, are visible, the Chinooks are said to have traditions of earthquakes that have shaken their houses and raised the ground (Clark, 1955).

   The Tolowa Indians of northern California and Oregon have a legend referring to inundation by a wave of sea water:

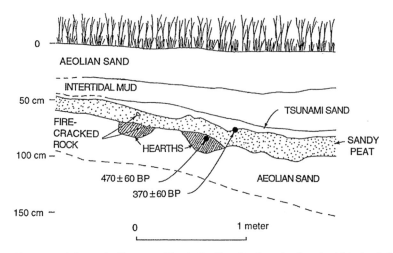

Figure 3.2 Schematic diagram of buried Indian fire hearths found within buried lowland soil on the Salmon River estuary. The buried soil is the result of sudden subsidence induced by a great earthquake about 300 years ago along the Cascadia subduction zone (redrawn from Minor and Grant, 1996).

> This happened in Oregon. There were no white people on earth when it happened. Chetko [Brookings] is where it happened . . . The two children ran in the house and told their grandmother, "Dog spoke," [an earthquake prediction?] . . . The grandmother told the children to go right away . . . as fast as they could, upstream away from the harbor toward Mount Emilie [Emily] . . . Halfway there they looked back . . . They could see the water come . . . They went back [the next day] to where their house had been . . . Everything was swept away clean . . . (Dubois, 1932, p. 261)

The Yurok Indians, who inhabited a region in northernmost coastal California, roughly coinciding with Redwood National Park, have many stories where the god *Yewol* (Earthquake) played a prominent role in many word-of-mouth, albeit allegorical, stories. Earthquake was a very powerful being who resided at Kenek and shook and tore the ground, disturbing trees, rivers, and even the ocean. Earthquake is reputed to have shaken trees at Espeu and Osegen and shaken the ground at Omen and Kohpei on the Pacific coast (Figure 3.3). He has also been described as trying to move the earth together at Rekwoi (Requa) so that he could cross the Klamath River. The locations are real but no longer extant and even though there is no factual basis for specific occurrences there is no doubt that the Yuroks felt earthquakes from offshore epicenters associated with the Cascadia subduction zone.

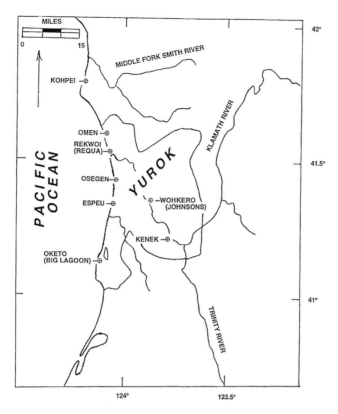

**Figure 3.3** Lands occupied by the Yurok Indians of the north coast of California. Locations of places described in myths are shown.

A documented story (Kroeber, 1976), told by a native, Dick of Wohkero, in 1907, is evidence that the Yuroks were aware of earthquakes:

> Now Earthquake is angry because the Americans have bought up Indian treasures and formulas and taken them away to San Francisco to keep. He knew that, so he tore the ground up there [referring to the earthquake of the year before in 1906; the upper end of the line of coast by Yurok reckoning is down on our maps].

In the Pacific northwest lies the Cascade Range, which includes the lofty volcanic peaks of Mount Rainier, Mount Adams, Mount Hood, Mount Shasta, and Crater Lake, the remnant of the ancient beheaded volcano Mount Mazama. Indians, such as the Klamath and the Modocs of Oregon and northern California who occupied the region for thousands of years before the arrival of the early pioneers, must have witnessed countless eruptions. Approximately 7000 years ago, as indicated by [14]C dating of charred tree stumps, the explosive decapitation and collapse of Mount Mazama, which formed Crater Lake, was probably preceded by enhanced earthquake activity, undoubtedly alarming the Klamath Indians who probably withdrew to a safer distance. Sandals and other artifacts have been found in the volcanic ash that blanketed the region for many miles. The story given below, which was passed on by word of mouth by the Klamath Indians, is worthy of note.

The Chief of the Below World sometimes came up from his place inside the earth and stood upon the top of the mountain that used to be (Mount Mazama). At one time he saw the beautiful maiden Loha whom he fell in love with. She rejected the advances of the Chief of the Below World who swore revenge on her people from the top of his mountain. The Chief of the Above World was seen from his position atop Mount Shasta about 160 km to the south. A furious battle began between the two spirit chiefs who shook the mountains and hurled rocks and flames at each other. An ocean of flame (a *nuée ardente* or glowing avalanche) devoured the trees on the mountains and the valleys prompting the call for a human sacrifice to placate the Chief of the Below World. Two old and revered medicine men lit their torches, climbed the mountain in view of the Chief of the Below World and jumped into the fiery pit below. The Chief of the Above World witnessed the brave act and as a result the earth shook so much that the Chief of the Below World retreated into his home inside the earth and the top of the mountain fell in upon him. Rain then fell for many years filling the great hole (Crater Lake). Removing from this tale all of its spiritual embellishments, it rather accurately depicts the mechanism of how Crater Lake originated and gives credence to the view that legends are often based in truth.

Earthquake occurrences are pervasive in the creation myths of the Zuñi Indians of the American southwest (Cushing, 1896). Their story first speaks of the instability of the world when it was still young and how "earthquakes shook the world and rent it." As a result the Zuñi people became wanderers, always moving east in search of a safer and more stable place, the middle of the world. In their search for the safety of the middle world the account relates 14 earthquake events and 15 stopping places before reaching their goal. The place of the middle was called Hálona Ítiwana, the site of the present pueblo of Zuñi in western New Mexico. Details of the Zuñi origin and migrations are perhaps only symbolic or at best open to various interpretations. It is known, however, that the Zuñi came from at least as far west as the Mojave Desert in eastern California, finally settling at the Zuñi pueblo around AD 1350. Moderate-sized earthquakes occur from time to time in eastern Arizona, which indicates that the early Zuñi travelers could have experienced such events (Figure 3.4).

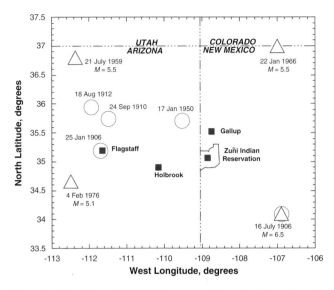

**Figure 3.4** Location of Zuñi Indian reservation and felt earthquake epicenters in the vicinity. Epicenters are listed in Table 10.1.

## 3.2    Mytho-history of the Aztecs, Maya, and Incas

One creation epic, the legend of the five suns, which has survived in the Nahuatl language of the Aztecs, has come down to the present time. In the era of the first sun *Naui Oceclotl* (Four Ocelot) the world was inhabited by people who were giants. *Tezcatlipoca* was the sun. At the end the people were eaten by jaguars and the sun itself was "destroyed." *Quetzalcoatl* knocked the jaguar out of the sky and started the time of the second sun, *Naui Eecatl* (Four Wind). People, houses, trees, and the sun itself were subsequently blown away by a terrible hurricane. A few survivors were turned into monkeys. The rain god *Tlalcanteuctly* became the third sun and started the era of *Naui Quiauitl* (Four Rain). This era was terminated when the people were destroyed by fire raining down from the sky from volcanic eruptions.

*Chalchiúhtlicue*, the wife of the rain god (also known as the Aztec patron of rivers, floods, and oceans) became the fourth sun, *Naui Atl*. Her tenure ended when the skies came falling down in the form of a world flood or deluge. The fifth sun, *Tonatiuh*, which exists today is *Naui Olin* or Four Movement:

> They say the sun that exists today was born in 13 Acatl [Reed or Cane, 751 AD) and it was then that light came, and it dawned. Movement Sun, which exists today, has the day sign Four Movement, and this sun is the fifth there is. In its time there will be earthquakes, famine . . . and because of this we will be destroyed. (Bierhorst, 1992b)

The Maya also believe in a creation myth where one world is destroyed and another created. The *Popul Vuh* (Recinos, 1950; Tedlock, 1985) of the Quiché Maya of the Guatemala highlands records a version of the creation depicting a mythical–historical sense of narrative time that includes a sequence of cycles that is still meaningful to the Maya today. The *Popul Vuh* manuscript, presumably written in the sixteenth century by a Quiché Maya scribe, consists of a preamble and four parts. In the first part the cosmic background of the different human beings created by the gods is told. The story relates how there was a man called *Vukub-Cakix* (Seven Macaws) who had two sons *Zipacná* and *Cabrakán*. *Zipacná* made the earth by shaping or piling up mountains. *Cabrakán* was the destroyer who could move mountains and the sky at will and make the earth tremble by tapping it with his feet.

That earthquakes play a role in world destruction is a theme to be found in Maya thinking. The Lacandon Maya, located in the forests of the Usumacinta Valley in eastern Chiapas, Mexico, were believed by many to be closest in bearing, physical appearance, and mannerisms to the ancient Maya. In the mid 1930s it was estimated that there were about 100 Lacandones left and that they were gradually dying out (Amram, 1937). One version of a creation myth related by the Lacandones involves two brothers, *Sukukyum* and his younger and less powerful brother *Nohotsakyum*. The younger brother is ordered to make a house for Sukukyum who lives in the middle of the world. Nohotsakyum made the world and everything in it, including the sun, moon, stars, plants for food, and finally people. Where Sukukyum lives there is much fire and he is capable of triggering earthquakes and volcanoes. Evil persons who kill, lie, and steal go down to where he lives after their deaths. His brother-in-law is *Cisin*. Cisin is a Lacandon deity, the lord of the underworld, who is also an earthquake god. He is a devil who lives underground in an ugly house and is always very angry at being shut up in the middle of the world. When he is angry he makes earthquakes by beating pillars that support the world (Cline, 1944; Thompson, 1970). He lives with jaguars who will end the world by eating the sun and the moon, making the earth black and cold. Cisin is also believed by some scholars to be *Ah Puch*, God A of pre-conquest Maya codices, a "God of Death." In the Dresden and Madrid codices he is pictured with a skull for a head, truncated nose, grinning teeth, bare ribs, and spiny vertebral projections.

A recurrent theme in the conversation of the Lacandones is the end of the world. At the end of the world all of the remaining Lacandon will gather at Yaxchilan. Yaxchilan is a Maya center located on a loop of the Usumacinta River along the Guatemalan border to the Mexican state of Chiapas. The Lacandones firmly believe that they are the only individuals able to pacify the ancient beneficial gods. With their demise the sun and the moon and other gods will fall

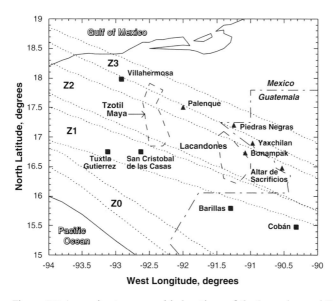

**Figure 3.5** Approximate geographic locations of the Lacandon and Tzotzil Maya. The dotted bands Z0 to Z3 show the locations of seismically active fracture zones induced by subduction at the Pacific Ocean convergent boundary. The location of fracture zones is taken from Vanek *et al.* (2000).

to the earth and the land will be destroyed by wind and earthquake. The wind is stated to be so strong that monkeys will be dislodged out of the trees. Myths notwithstanding, there is little doubt that because of their geographical location the Lacandones undoubtedly felt earthquakes (Figure 3.5). It should not be at all unusual to find traditions of natural disasters amongst the inhabitants of these regions.

The Tzotzil Maya occupy a region of the highlands of Chiapas, Mexico. One of their beliefs is that four gods sustain the world on their shoulders. These gods are known as *Cuch Uinahel Balumil*, who are the sky and earth bearers that are set at the far corners of the world. Any slight movement by these gods will produce a tremor of the earth or an earthquake. Gods of the four familiar cardinal points are positioned at locations intermediate between those of the Cuch Uinahel Balumil (Guiteras, 1961; Holland, 1963).

The Chorti Maya (Wisdom, 1940) are positioned in an area of southeastern Guatemala roughly bordered by 89° to 89.5° W longitude and 14.5° to 15° N latitude straddling the border with Honduras (Figure 6.15). Copán was their cultural center. Their proximity to the Motagua and Chamelecon faults, within the North American–Caribbean plate boundary zone, has meant that the area is continually subjected to earthquakes and as a result has played a role in the mythology of the Chorti. In Chorti belief there are sky *Chicchans* who are four giant serpents or snakes, each of whom dwell at the bottom of a large body of water at the four cardinal directions. The sky Chicchans can produce violent storms, lightning, and cloudbursts. There are, in addition, many earth Chicchans who reside in streams, springs, and lakes and travel upstream during the dry season to reside in springs in the hills. With the beginning of the rainy season they re-enter the streams and at times produce flooding. When the Chicchans move within the hills by turning in their sleep they produce earthquakes. A severe earthquake takes place if the Chicchan decides to turn over completely and lie on his opposite side. Interestingly, Chicchan, represented by a cross-hatched spot on the eyebrow or temple, is the fifth day sign in the Maya calendar.

The use of the "sun" to describe the chronology of the five ages of the world was not solely restricted to the Aztecs. It was also a common practice for the Incas of the Andes:

> since the creation of the world until this time, there have passed four suns without [counting] the one which presently illumines us. The first was lost by water, the second by the falling of the sky on the earth . . . the third sun they say failed by fire. The fourth by air: they take this fifth sun greatly into account and have it painted and symbolized in the temple Curicancha [the Inca Temple of the Sun in Cuzco] and placed in their *quipus* [knotted cords used for counting and record-keeping until the year 1554]. (Sullivan, 1996, p. 27)

Time in the Aztec Nahuatl calendar was bundled in 52-year rounds. This cycle makes the determination of absolute time increasingly artificial when extrapolated backward to date events prior to the early 1500s. The arrival of Cortés in AD 1519 corresponds to 1 Acatl in the ritual calendar of the Aztecs; AD 1479 is equivalent to 13 Acatl in the reign of Ayaycatl. Placing an absolute date on the earthquake occurrence described below may have some literary value but must be viewed with caution for its historical value:

> Cumplidos ciento cincuenta y ocho años después del gran huracán, y cuatro mil novecientos noventa y cuatro de la creación del mundo, tuvieron otra destrucción los de esta tierra que fueron los quinametin, gigantes que vivían en esta rinconada, que se dice ahora Nueva España, la cual destrucción fue de gran temblor de tierra, que los tragó y mató, reventando los altos montes volcanes, de suerte que se destruyeron todos sin escapar ninguno, y si escapó alguno fue de los que estaban más hacia la tierra dentro; y asimismo muchos de los tultecas murieron y los chichimecas sus circunvecinos, que fue en el año de ce tecpatl; y a esta edad le llamaron *Tlacchitonatiuh*, que quiere decir sol de tierra. (Alva Ixtlixóchitl, 1975, I: 264–265)

The quotation describes a gigantic and tragic earthquake in New Spain in the year 1 Pedernal (Flint, Stone Knife, Tecpatl) that leveled mountains. It is stated that the event took place 158 years after the great hurricane and 4994 years after the creation of the earth and that many Toltecs were killed. It is interesting to speculate on a date. The beginning of the Toltec era lies somewhere in the time interval from AD 694 to 751 and it is known that the Toltecs inaugurated their first ruler in 1 Pedernal (AD 752). The next earliest written record of an earthquake in Mexico is found in the *Anales de Tlatelolco* (1948, p. 56): "En el año 13 Calli [Casa, House, AD 1453] . . . hubo también terremoto . . ." Prior to 13 Calli the closest date of 1 Pedernal would be AD 1428, placing the described event as taking place on one of the 52-year cycles between AD 804 and 1428, and the creation of the earth sometime between 3566 BC and 4190 BC. For comparison, time as reckoned in the Maya Long-Count calendar tabulates the days from a "zero" or starting date of 3114 BC. Traditionally, the Jewish calendar places the creation of the earth at 3761 BC; christian scholars would place it at 4004 BC. The inference about earthquakes appears clear; however, one must not be trapped into a web of mysticism, attempting to read into simple statements meanings more precise than those the Aztecs themselves might have formulated.

The rumbling and shaking of the earth has in some instances been mentioned in association with human ceremonial sacrifice. One type of Mesoamerican sacrifice involved binding a victim to an upright wooden form or scaffold and then killing them by darts and arrows (Landa, 1938; Recinos *et al.*, 1953; Durán, 1971). The place of execution where the scaffold sacrifice to *Toci*, the Aztec earth goddess, is honored during the *Ochpanitzli* ceremony (the 12th of the eighteen 20-day months of the ritual calendar), is believed to tremble and shake. The Cakchiquels, a Maya tribe in the Guatemalan interior, allude to the trembling and shaking of the earth at the place where the sacrificial execution of *Tolgom*, the "sun of the mud that quivers" took place.

It is true that many beliefs and stories reflect real historic content. Many of the legends, which probably encompass a collection of traditions, are highly imaginative. It would be misleading to attempt an explanation based on one earthquake or volcanic eruption occurrence alone or extrapolate the meaning of one local tradition alone to a much wider geographic area. Nevertheless, the recollections of specific events are often consistent with local geological and tectonic conditions. This in itself is not surprising. More than 80 percent of the energy released in the world's earthquakes takes place in the circum-Pacific belt. Volcanoes are associated with this zone of seismicity – hence the term "Pacific ring of fire." Tsunami, triggered by circum-Pacific zone earthquakes, undoubtedly gave rise to many flood tradition stories along the Pacific coast of the Americas since flood deluge stories are often associated with, and can be attributed to, a temporary rise in sea level. It is not at all unlikely, therefore, that disaster stories are universal in the traditions indigenous to their localities. These tales have been embellished by story tellers and shamans and passed on from generation to generation with a religious or symbolic message to relate. Many natural disaster stories have a common thread because earthquakes, volcanic eruptions, and floods are universal geological catastrophes that frequently occurred and will continue to occur along the rim of the Americas.

Another aspect of earthquakes that is of more than passing interest is the subtle psychological effect on the mind-set of individuals continually subjected to such catastrophes. Chile has experienced a large number of earthquakes over the past 450 years or so. Some historians put the count at one damaging event every 2 to 3 years and one tremor felt somewhere, on average, every other day. Not all of these events are major disasters but they have had their effect on the minds of the Chileans. It is said that Chileans are imbued with a mysticism about earthquakes that goes well beyond fatalism. Earthquake damage is often routinely understated. If earthquakes are Chile's special purgatory, it was aptly described by Albert Camus (1987, p. 133) as "a psychology of instability." What is the loss of worldly possessions when life itself is a gamble?

4

# Earthquake effects

Earthquake effects can be classified as primary, transient, or secondary, although the rigorous separation of effects is not always possible unless they are documented during an earthquake. Nevertheless, primary effects are the permanent geological features produced by the earthquake that leave a lasting and sometimes indelible visual imprint. Among these features are surface ruptures, scarps, ground elevation changes, and offsets of natural and man-made structures. Primary effects lead to transient effects – mainly ground shaking, sandblows, liquefaction, and the generation of tsunami. It is the transient effects that produce secondary effects, such as landslides, ground slumping, damage to buildings and other structures, and secondary fires. Ground shaking is the principal cause of damage from earthquakes.

## 4.1  Seismic intensities and ground accelerations

Prior to instrumental observations, it was realized that the effects of an earthquake could be quantified using a numerical scale that specifies the apparent intensity of the tremors or the level of shaking. These data, presented as isoseismal maps, are of particular importance when available, because they provide data for analyzing the effects of destructive earthquakes of the past. Two intensity scales have been used for the study of early American earthquakes, the Rossi–Forel Seismic Intensity Scale and the Modified Mercalli Intensity Scale. The Rossi–Forel Scale is no longer in use but values from this scale can be converted, with some uncertainty, to Mercalli intensity values (Appendix C). The Modified Mercalli Intensity Scale has numerical values ranging from I, being virtually indiscernible or undetectable without instruments, to XII, producing total destruction of most buildings. In this scale the lower intensity values of I to V delineate

Table 4.1. *Seismic intensity, Modified Mercalli Scale*[a]

   I  Not felt except by a few under especially favorable conditions. Best considered as an instrumental shock, noted by seismic instruments only. Accelerations less than 1 cm/s$^2$ (0.001 g).

  II  Felt indoors by a few, especially those at rest on upper floors of buildings or by sensitive and nervous people. Delicately suspended objects may swing. Some birds and animals reported to be uneasy or disturbed.

 III  Commonly felt indoors but may not be immediately recognized as an earthquake until it was related that others had felt it. Standing automobiles may rock slightly as if from vibration caused by a passing truck. Duration may be estimated. Rare reports of being frightened. Accelerations ranging from 0.17% to 1.4% g.

 IV  In daytime felt by many indoors but by only a few outdoors. Awakens some light sleepers. Distinctive rattling of dishes, windows, doors, and creaking of walls. Sensation like a heavy truck passing. Standing automobiles rocked considerably. Nothing is knocked over or falls from shelves. Accelerations between 1% and 4% g.

  V  Felt indoors by all, and almost everyone outdoors. Awakens most people, frightens many. Some dishes and window glasses broken, wall plaster may crack. Reports of buildings trembling/groaning, doors swinging open and closed. Disturbance of telephone poles, trees, and tall objects sometimes noticed. Pendulum clocks stop. Hanging pictures knock against walls or swing out of place. Accelerations about 4% to 9% g.

 VI  Felt by all, awakens all, frightens most people. Many run outdoors. Many report difficulty in standing or walking, and duck and take cover. Heavy furniture may be moved, plaster falls. Some damage to buildings not adequately tied to foundations. Damage and some toppling of unreinforced masonry chimneys. No structural damage to well-built structures. Adobe structures cracked. Can be noticed by people driving slowly. Some reports of trees falling, and particularly ripe fruit, dead branches, and tops. Some reports of liquefaction effects such as sand blows in water-saturated, unconsolidated areas. Horizontal accelerations of 9% to 18% g.

VII  Everyone runs outdoors. Difficult to stand. Felt in moving automobiles. Frightens all, general alarm. Damage to masonry structures. Most unreinforced masonry chimneys damaged, some broken at roof-line and knocked down. Fall of plaster, loose bricks, stones, tiles, cornices, unbraced parapets, and architectural ornaments. Frequent falling of branches and treetops in heavily wooded areas. Rockfalls and landslides in steep areas. Waves on ponds and water turbid with mud. Reports of foundation damage to poor structures. Some cracks in masonry of ordinary workmanship and mortar. Some tilting and toppling of gravestones, monuments, and stelae. Horizontal ground accelerations ranging from 18% to 34% g.

VIII  Frightens everyone, many people panic. All report difficulty in standing with reports of people falling and being knocked over. Hinders driving of automobiles. Major damage to unreinforced masonry structures and partial collapse. Fall of stucco and some masonry walls. Wooden-framed houses moved on foundations if not bolted down, loose panel walls thrown out. Some damage even in buildings of good design and construction. Cracked and disturbed ground may be common. Changes in flow of springs and wells. Landslides in steep areas and dust ejected into air if ground dry. Trees shaken strongly – branches and trunks broken off, especially palm trees. Twisting and falling of chimneys, columns, and monuments. Ground accelerations of 34% to 65% g. (Decidedly unpleasant human sensation.)

IX General panic. Adobe structures destroyed. Heavy damage, sometimes with complete collapse to ordinary, unreinforced masonry structures. Serious damage (particularly foundations) to reinforced masonry structures. Some well-built wooden-framed houses suffer structural damage. Serious damage to reservoirs. Conspicuous cracks in ground. In alluviated areas ejection of sand and mud. Liquefaction. Accelerations of 65% to 124% g.

X Wooden houses of good design and construction collapse. Most masonry and frame structures destroyed together with their foundations. Ground cracks causing damage. Rails are bent. Slopes and embankments slide. Sand and mud shifts horizontally on beaches and flat land. Dams are seriously damaged. Water is thrown on banks of canals, lakes, and rivers.

XI Few masonry structures remain standing. Bridges destroyed. Fissures over large areas of ground. Underground pipelines completely out of service. Rails bent prominently. Earth slumps and land slips in soft ground.

XII Damage is total and practically all works of construction are damaged or destroyed. Lakes are dammed, new waterfalls are produced, and river courses are deflected. Lines of sight and level are distorted. Widespread and common incidence of objects being thrown into the air.

---

[a]Based on scale presented by Wood and Neumann (1931) and Richter (1958a) with useful diagnostic descriptions added by author, and Dengler and McPherson (1993). Ranges of ground accelerations are based on combined regression of peak accelerations versus observed intensity for a number of significant California earthquakes (Wald *et al.*, 1999).

levels of shaking that are felt by individuals. Values greater than V describe levels of damage noted in buildings and other structures. Principal shortcomings and uncertainties in assessing seismic intensity are local geological site effects, soil liquefaction, population bias, and the descriptive exaggeration by people that have not experienced earthquakes. Even with its shortcomings and uncertainties, seismic intensity is often the only parameter available from which to quantify the level of ground shaking at a particular locality following damaging earthquakes. Table 4.1 lists the scale, together with annotated descriptions applicable to the different levels of intensity.

Intensity VI is the level at which structures built of weak materials such as adobe, or those bonded with poor-quality mortar, begin to be damaged. Intensity levels of VII and VIII produce damage in structures that use good-quality bonding materials and workmanship. At these levels of shaking, one would expect to see damage effects preserved in the archaeological record. Uncertainties can arise in intensity estimations, however, because levels observed at a site depend on its azimuth and distance from the earthquake epicenter. Levels of intensity can also be amplified by local soil conditions and by topographic variations of the ground. Larger earthquakes also radiate seismic energy along a finite fault rupture such that focusing and defocusing effects, analogous to the Doppler effect, can modify the level of shaking observed.

Intensities of XI and XII are difficult to distinguish and are of very rare practical importance. XII means the maximum conceivable effect and may not even be reached during an earthquake. In common usage, therefore, the Modified Mercalli Intensity Scale is in fact a ten-degree scale. Earthquakes also produce effects on the ground that are included in the intensity scale. These seismo-geological manifestations include hydrological effects, slope failure effects, and surface ground rupture and breakage. Care must be taken when using these features to assign intensity values. Ground ruptures, indicative of intensities of at least VIII to IX, can be confused with fissures caused by ground shaking. True landslides, from dislodged rocks produced by ground shaking, may be confused with features produced by man-made unstable slope conditions. Hydrological effects such as liquefaction are usually indicative of an intensity level of IX.

Buildings and structures are not designed to withstand surface fault ruptures. Most structural damage is due to the level of horizontal acceleration that is applied to the base of the structure. The vertical component of ground acceleration is of less importance since structures are designed to stand up under their own weight. Ground accelerations that are applied to a structure are specified as a percentage of the Earth's gravitational acceleration, g; $1\,g = 980$ cm/s$^2$ (32 ft/s$^2$).

There is no fundamental reason to expect that a simple relationship exists between a Modified Mercalli intensity value and a specific ground motion parameter such as ground displacement, velocity, or acceleration. Nevertheless, one common correlation in use is horizontal peak ground acceleration recorded by instruments versus the Modified Mercalli intensity value determined. Peak ground acceleration does play a central role in the design aspects of ground motion. A frequency-versus-acceleration amplitude spectrum is used in earthquake engineering design and the value of the spectrum at the high-frequency end is scaled by using the value of the maximum peak ground acceleration. Some regressions based on peak ground motions and intensities observed for earthquakes in California and Mexico are shown in Figure 4.1.

The linear regressions show that as more instrumentally recorded strong ground motion data have become available, lower values of seismic intensity are indicated for a given level of ground acceleration. Conversely this indicates that higher levels of peak ground acceleration would be inferred for a specified value of Mercalli intensity. An intensity value of X suggests ground accelerations approaching 100% g. Acceleration levels less than 1% g would correspond to

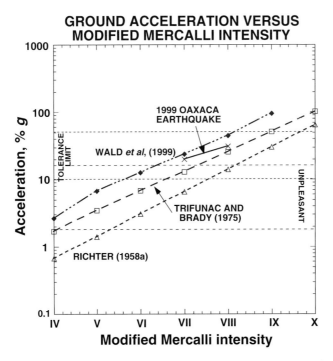

**Figure 4.1** A plot of the logarithm of horizontal ground acceleration (% g) recorded during earthquakes versus the Modified Mercalli intensity at the recording site. The lines of Trifunac and Brady (1975) and Wald *et al.* (1999) are based on southern California earthquake data. X indicates values measured during the 1999 Oaxaca, Mexico, earthquake.

intensity levels of I to III that are difficult for most individuals to separate. Seismic intensities of V or greater correspond to accelerations in excess of ∼3% g. Depending on the sensitivity of a particular individual, this level of acceleration can begin to cause discomfort. Intensity levels of VI to VIII encompass ground accelerations in the range of 10% to 60% g. At 10% g, ordinary structures that are not specifically designed to be earthquake resistant, will be damaged.

## 4.2    Fundamental principles of earthquake-resistant design

When a building is violently oscillated, walls can shear, beams can buckle and fail, and columns can compress and collapse. The effects of earthquake shaking are studied by applying lateral forces that simulate the inertial loads that a structure is subjected to when it is oscillated, or accelerated from side to side, during an earthquake. Lessons learned from observing the damage produced by many earthquakes over the past 100 years have emphasized

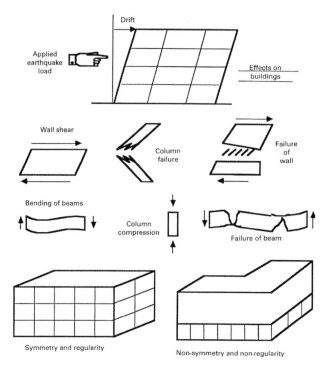

**Figure 4.2** Types of failure when a building is subjected to horizontal accelerations and the basic principle of earthquake-resistant design.

a number of features that are pertinent when one is searching for past earthquake effects at archaeological sites.

Adobe structures and walls formed of rubble hearting or lime-mortar rubble possess a low tensile resistance and are not safe from the effects of earthquake shaking. Veneers of stone are susceptible to damage unless they are thoroughly anchored to a fairly rigid backing. Structures with towers and other roof appendages are at a great disadvantage during an earthquake. The shape of the floor plan of the building, whether it is symmetrical and rectangular, or L- or E-shaped, affects the resistance to ground shaking. Uniformity in the type of construction with height, as opposed to irregularity in building materials, is a desired attribute.

The two overriding keys to earthquake-resistant construction are symmetry and regularity with respect to height (Figure 4.2). Structures with drastic irregularities in the floor plan will experience greater damage from tremors than those that are regular and symmetrical in shape. Architectural ornamentation and veneers add little intrinsic strength to a building.

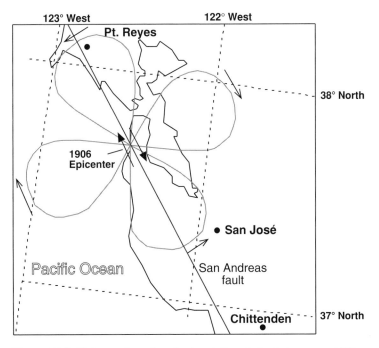

**Figure 4.3** Radiation pattern for a horizontally polarized shear wave (SH) propagating to the northwest from the 1906 San Francisco earthquake epicenter.

## 4.3    Directions of fall of monuments and tombstones

Columns, monuments, and gravestones are often toppled during earthquake shaking and typically fall in the direction of the main ground motion caused by the tremor. Engineers and seismologists agree that the majority of damage features produced by earthquakes are the result of horizontal disturbances produced from transverse or shear waves. The concept of a preferential direction of fall can be illuminated using features noted during, and shortly after the 1906 San Francisco earthquake.

Current opinion places the epicenter of the 1906 shock on the coast about 10 km south of San Francisco (Figure 4.3). During the earthquake shaking the conductor on a train leaving southeasterly from the Pt. Reyes station, 60 km northwest of the epicenter, described feeling a lurch to the east (indicating an initial ground motion to the southwest) followed by another lurch to the west (ground motion moving back to the northeast), which toppled the entire train on its side. The final direction of the fall of the train indicates that the main displacement pulse acted in the southwest direction. A pulse of displacement acting in the southwest direction is consistent with the radiation pattern for a horizontally polarized shear wave (SH), propagating to the northwest by rupture, approaching the Pt. Reyes station from the epicenter on the San Andreas fault located to the southeast. The 1906 earthquake rupture on the San Andreas fault was bilateral so that locations southeast of the epicenter (rupture approaching from the northwest) would expect a displacement pulse in the northeast direction; this is exactly what was observed. Tombstones in San José and Chittenden were thrown east-northeasterly. Lines drawn at right angles to the more common direction of fall will represent the azimuth from the main source of radiated seismic energy.

Free-standing columns, formed by the stacking of individual cylindrical drums or rectangular shaped blocks, often supported a roof structure. Inertial effects would cause the roof to move in the opposite direction to the initial direction of strong ground motion. The base of the columns acts as a hinge when failure and collapse begins, resulting in neat rows of fallen columns in a domino-style arrangement covered by the main mass of the roof.

### 4.4    Recent earthquake damage at Mitla and Monte Albán

The peoples who inhabited the Oaxaca area of Mexico, and those living there today, are no strangers to earthquakes. The constructions at Mitla and Monte Albán undoubtedly were modified and evolved over time because of their location in a zone of frequent earthquake activity. Their marked horizontal tendency exhibits a formal obedience to the principles of symmetry and regularity, and awareness of the continuing menace of earthquakes. The search for evidence of historic earthquake damage at these sites, is difficult however, because the areas were in ruins or were put to other uses prior to the Spanish conquest. In addition, the invasion of the Spaniards ushered in an era of intolerance and the systematic destruction of many ancient cultural sites, such as Mitla.

Most of the structures that are seen today at Mitla were built in the thirteenth and fourteenth centuries. The complex of structures includes the Group of the Columns, the Arroyo Group, and the Northern Group (Figure 4.4). The Parish Group, also referred to as the Northern Group, had a church built in 1590 on top of one of the structures at Mitla using some of the original elements as construction material. Portions of the complex forming the Group of Columns managed to survive because of the thickness of the walls, the lack of windows in the walls, and the alternating, interlocking pattern of blocks used in the wall construction. The bases of the structures leaned inward offering the partial stability of a pyramid.

A clever solution to the threat of earth tremors is found at Mitla. The roofs of tombs use large, regular slabs of stone laid without mortar. A recessed row of small, loose stones is placed between the slabs and the top of the walls; the stones

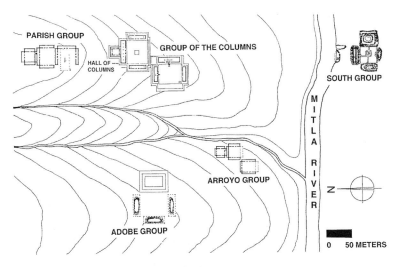

**Figure 4.4** Site layout of the ruins of Mitla.

act as ball bearings giving a form of flexibility to the structure by allowing it to absorb the seismic shocks from earthquakes. Even with these clever additions, earthquake damage continues to affect these archaeological sites.

On January 15, 1931, an $M = 7.8$ earthquake struck the Oaxaca area causing great destruction in the city of Oaxaca. About one-third of the seismic activity of Mexico takes place along the coastal region of Oaxaca. Earthquakes here are the result of the subduction of the Cocos plate beneath the North American continental plate. The isoseismal map (Figure 4.5) shows that an intensity of IX was reached within the meizoseismal area. On September 30, 1999, the Oaxaca area was struck again by an $M_w = 7.5$ earthquake whose hypocenter was close to that of the 1931 event. The isoseismal map for the 1999 event is shown in Figure 4.6. These maps emphasize that the archaeological sites of Monte Albán and Mitla have been subjected to intensity levels of VII to VIII. Several digital, strong-motion instruments recorded the 1999 event and accelerations of 20% g were observed at Oaxaca at a distance of ∼100 km from the hypocenter.

Archaeological investigations at Mitla and Monte Albán that were begun in the early half of the twentieth century did not focus on documenting possible earthquake damage effects of the past. As a result, reconstruction and repairing undoubtedly masked the damage from earlier earthquakes. What types of damage were produced at Mitla and Monte Albán from the 1999 event? Let us look at the south (front) face of the Hall of Columns at Mitla. The front and interior façades of the hall consist of panels of stepped fret designs. Stones were cut with great precision and inlaid in a mosaic fashion into the core of the wall to form a "frozen-lace" design. The principal damage caused by the 1999 earthquake was

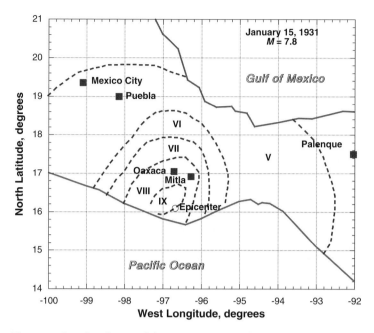

**Figure 4.5** Isoseismal map of the 1931 Oaxaca earthquake.

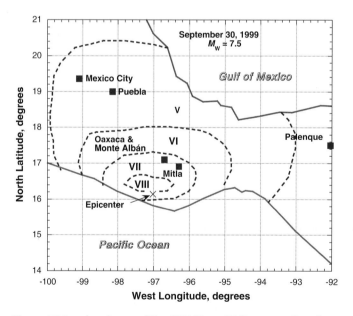

**Figure 4.6** Isoseismal map of the 1999 $M_w$ = 7.5 Oaxaca earthquake.

Figure 4.7 Lateral separation of the panels on the south wall of the Hall of
Columns at Mitla produced by the 1999 event.

lateral separation of many of the precisely fitted vertical joints along the walls
of the Hall of Columns (Figure 4.7). The missing portions of the stone mosaics
fell from the walls at an unknown earlier date (a photograph taken by Désiré
Charnay in 1859, republished in 1994, shows the same missing panels).

Monte Albán, perched on the top of a flattened mountain above the city of
Oaxaca, is one of the most extensively studied archaeological sites in Mexico
(Figure 4.8). Restoration and excavation at this site have been intermittently

Monte Albán

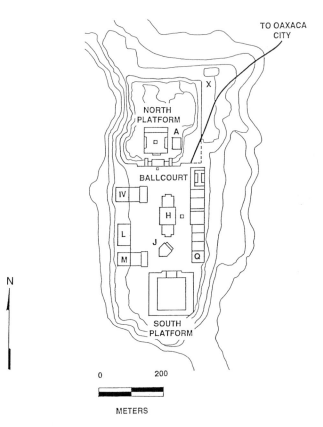

Figure 4.8 Site map of the structures at Monte Albán.

under way since the 1930s. The earthquake of September 30, 1999, damaged more than 20 structures in the site complex. A number of structures were severely affected and many of the remaining buildings suffered minor damage to their bases and façades. Some examples of the damage produced by the 1999 event are shown in Figures 4.9 to 4.14. Examination of the recent damage is useful because it gives clues as to the types of features to search for as evidence of earlier earthquakes.

The main damage features produced by the 1999 Oaxaca earthquake can be summarized as follows:

(a)   collapse of walls;
(b)   collapse of reconstructions and excavations;
(c)   cracks and failures in walls, floors, tableros, and basement platforms;
(d)   tilting and loss of plane levels;
(e)   cracking of interior wall murals.

**Figure 4.9** Damage to the temple just south of Ballcourt at Monte Albán. The 1999 Oaxaca earthquake produced a horizontal crack at the base of a low platform.

**Figure 4.10** Staircase at the south end of Building H damaged in the 1999 Oaxaca earthquake. A pronounced jagged vertical crack can be seen.

**Figure 4.11** Scaffolding erected after the 1999 Oaxaca earthquake for stabilization of the southwest stairway to Ballcourt at Monte Albán.

**Figure 4.12** Close-up view of southwest stairway to Ballcourt. A jagged crack can be seen running vertically between the bracing.

Figure 4.13 Collapse produced on the northeast side of Mound Q at Monte Albán by the 1999 Oaxaca earthquake. Mound Q was in the process of being excavated.

Figure 4.14 Earthquake-produced damage to the northeast side of the front upper wall of Structure IV at Monte Albán. Note the prominent crack in the center of the image. The scaffolding was erected after the 1999 Oaxaca earthquake to prevent further damage.

### 4.5    Maya architecture and the corbeled arch

Given the fact that the Maya lived in lands subjected to earthquakes, were they aware of any rudimentary principles of earthquake-resistant design for their structures? Before attempting to answer this question a few general comments about the construction techniques of the Maya are in order. The Maya stonemasons were ignorant of the true arch, which is bridged by a single capstone, the "keystone" that anchors the whole arch. Instead, buildings were roofed by a series of overlapping stones that approached one another until they could be bridged with one or a series of single slabs. The two halves of the arch are structurally independent and care must be exercised in keeping everything in balance to achieve stability from the dead weight of the material. If the vault were not ideally centered, the center of gravity would act to overturn the overlapping stones if they were not held up by the other half of the vault. Above this false or "corbeled" arch, a thick layer of solid masonry was added. Corbeling is a form of cantilevering. The only support is from the overlap provided by the block below; each stone is counterbalanced by the weight above, which holds it in place. The principle is a *downward thrust of a load on overlapping stones*. The corbeled arch is much weaker than a true arch and requires great thickness of walls to withstand the outward thrust of the false arch. As a result the interior chambers of Maya buildings were of very restricted width, rarely exceeding 3 m.

Many investigators have praised the vaulted-roof system as being inherently sturdy compared to the true arch. The true arch will collapse if the keystone (or in reality any other segment) is removed. This statement may only be a half-truth. Many corbeled Maya arches have been found where one side has collapsed and the other half remains standing. In terms of earthquake stability, the Maya arch is vulnerable to lateral shaking since its strength depends on a vertical load and the proper positioning of the overlapping stones (Figure 4.15).

Many of the Maya doorways were bridged with wooden lintels whose subsequent decay has led to collapse of the doorway. Small building blocks were used with and without high-quality mortar and the rapid growth of tropical vegetation could easily thrust roots between the blocks, so contributing to the subsequent collapse of many Maya structures. With their use of small blocks the Maya were not as advanced as the megalithic stone builders of the Peruvian and Bolivian highlands.

Maya walls were often not solid masonry block walls but rather consisted of a single layer of masonry that served as a facing for a rubble hearting. For example, the walls at Copán have a core or hearting of pounded earth or clay mixed with broken stone. Such a hearting gives greater mass and weight to the

**Figure 4.15** A Maya corbeled arch. The right-hand side of the arch is close to separation and failure.

wall and also enables it to be used as a retaining wall for an earth embankment. If the masonry or stonework veneer on a mud-walled structure is washed away by heavy rains the structure is subsequently weakened. As a result they are more prone to collapse from earthquake action.

On the roof of many Maya structures a superstructure is found, a roof-comb, varying from a type of lattice work of separate walls leaning towards each other to a single wall resting on the top of a temple or platform that extends above one of the walls below. Roof-combs supported by the front wall are called "flying façades" because they give a false impression of the height of the building. Some researchers have suggested that roof-combs were placed on the top of structures

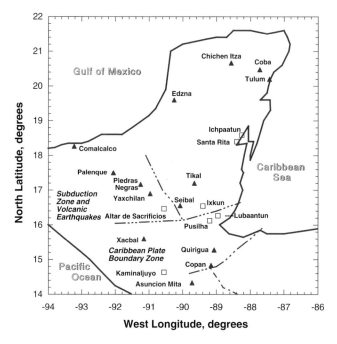

**Figure 4.16** Location of Maya sites with corbeled structures relative to the
Caribbean plate boundary seismic zone and the Pacific coast zone of subduction
and volcanic earthquakes. Open squares indicate sites where corbeled structures
are absent.

as additional weight to help support the vault beneath. This conjecture seems
contradictory to its evolution in that many roof-combs constructed later had
more open space and possessed less weight. It seems more likely that roof-combs
were ornamental or commemorative of some important personage. Whatever
their purpose, roof-combs are susceptible to vibrational damage from ground
shaking.

The corbeled arch in the realm of the Maya had its apparent origin in the
Petén district of Guatemala around AD 328. Its use expanded into most of the
areas that were occupied by the Maya by AD 889. However, there are Maya
regions where the corbeled vault is not found. Such vaults are not found at
Ixkun, Pusilhá, and Lubaantun, and are conspicuously absent in the highlands
of Chiapas and Guatemala, with the exception of small chambers at Asunción
Mita about 90 km south-southwest of Copán. The most westerly corbeled vault
site is at Comalcalco in the Mexican state of Tabasco (Figure 4.16). It is possible
that the frequent occurrence of earthquakes in the highlands may have influenced the architecture of the ancient Maya and advised them against the use of
the corbeled arch.

The corbeled masonry vault must have been the end product of an evolutionary process. Its predecessor was undoubtedly a wooden hut or a thatch-roofed building whose lower walls were constructed of vertical poles and horizontal cross-members forming the framework for a woven lattice of sticks; this lattice was plastered or packed on the inside and outside with mud, which was mixed with straw or another binder, to form a reinforced adobe wall about 1.5 m in height. This construction, known as *bajareque* (pronounced "bahareque"), is very earthquake resistant, provided that the wood has not been weakened by rot or termites. The structural resistance comes from strength and flexibility achieved by tying strong wall poles to the horizontal roofing members. Masonry walls can be thickened to enable the erection of a corbeled stone vault on top of walls that previously only supported a wood-and-thatch roof.

Aside from the corbeled arch, the inclusion of superstructures such as roof-combs and flying façades is common in regions less subject to earthquake activity. No roof-combs are found at Quiriguá and Copán, whereas at more northerly and less earthquake-prone sites such superstructures are common.

The typical Maya building is a solid box-like structure containing one or more narrow vaulted chambers. Some examples of Maya architecture are shown in Figure 4.17. All in all the buildings of the Maya possess two essential elements of earthquake-resistant construction: *regularity* and *symmetry* with height. Even today, structures that possess drastic irregularities in floor plan from story to story will experience greater damage from tremors than those that are regular and symmetrical in shape. The material that was used for many of the Maya structures was a high-quality limestone. In principle, many of the remaining buildings would have been found in an excellent state of preservation, but for human and natural causes. Of course implicit in this statement is the need for high-quality bonding material and well-braced walls. If the bonding material is absent or loose, airborne seeds can fall into open joints between stones or blocks and take root. The resulting vegetation can push its way between the stones forcing them apart. Structures weakened by shaking from earthquakes would be particularly susceptible to subsequent destruction from vegetation.

Evidence against vegetation being the sole cause of building failure and collapse has been found, however, at several Maya excavations. After the surficial growth has been cleared, many structures are in an excellent state of preservation arguing that collapse took place before any vegetation grew upon the structure. Many Maya structures have, because of their geographical location, been subjected to the effects of earthquakes. Regularity and symmetry in building design, whether intentional or not, were instrumental in protecting many of these structures from earthquake effects.

# Maya Architecture

Single-chambered building

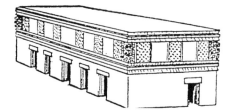

Multiple-chambered building

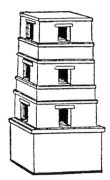

Square tower (restored)
at Palenque

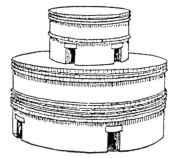

The round tower (restored) at Chichen Itza

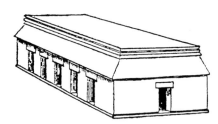

Building at Chichen Itza with sloping
entablature of Palenque type

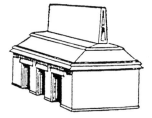

Palenque-type of temple

**Figure 4.17** Some examples of Maya architecture. With the exception of roof-combs, the Maya structures possess two key ingredients of earthquake resistance, i.e. symmetry and regularity.

Table 4.2. *Criteria for identifying earthquake occurrences from archaeological data*[a]

1. Ancient constructions offset horizontally or vertically by surface fault ruptures.
2. Skeletons of individuals killed and buried beneath the debris of fallen structures.
3. Abrupt geomorphological changes, such as changes in drainage and irrigation patterns associated with destructions and/or abandonment of buildings and sites.
4. Diagnostic structural damage and construction failures:
   - displaced drums of dry masonry columns;
   - opened vertical joints and parts of dry masonry walls slid horizontally;
   - diagonal cracks in rigid to semi-rigid walls;
   - inclined or subvertical cracks in the upper parts of arches, vaults, and domes, or their partial collapse along these cracks;
   - dropped keystones in dry masonry arches and vaults;
   - several parallel fallen columns;
   - several fallen columns with their drums in a domino-style ("slices of salami") arrangement;
   - construction deformed as by horizontal forces (rectangles transformed to parallelograms).
   - damage to dry masonry walls (tapiales), rock walls and other structures and evidence of hasty repair or patching (using different size stones or other construction materials);
   - collapse of corbeled roof vaults from wall separation;
   - cracks or separations in floors, tableros, and bases of platforms.
   - walls out of plumb;
   - deformation or pushing down of walls and floors.
5. Evidence of destruction and hasty reconstruction of sites, with the introduction of what might be regarded as "anti-seismic" building construction techniques, such as buttressing of adobe walls.
6. Preferentially tilted or fallen stelae.
7. Broken pottery found in fallen position.
8. Well-dated building destructions correlating with historical (sometimes epigraphic) evidence of earthquakes. Earthquakes mentioned in myths and legends.
9. Damage or destruction of isolated buildings or whole sites, for which an earthquake appears to be the only reasonable explanation.

[a] Adapted from Nur and Ron (1997a, b), and Nur and Cline (2000) with additions by the author.

## 4.6    Archaeological indicators of past earthquake occurrences

Earthquakes have undoubtedly caused damage to prehistoric sites and may have left evidence in the archaeological record. In searching for earthquake effects the problem should not be approached with a view to proving that an earthquake or a seismic catastrophe was the sole cause for the abandonment of a particular site. On the other hand, efforts should focus on whether *any* evidence for past earthquake damage can be found and what role earthquakes may have played in a sequence of debilitating events.

There are a number of diagnostic criteria that, if noticed at an ancient site, are suggestive of damage by an earthquake. These criteria include walls that have collapsed or have been patched and reinforced, crushed skeletons, and monuments or columns toppled in systematic and preferential directions. Many other suggestive criteria are summarized in Table 4.2. Obviously, care must be taken in attributing to earthquakes all the damage features noted.

Walls are found tilted, leaning, or with fallen sections at many sites. These features should not be summarily explained as the result of poor supporting soil conditions, slow non-earthquake ground movements, or uneven stresses in the wall set up by poor construction techniques. Evidence of rebuilding, often with different materials, and the addition of buttresses, are strong indicators of exposure to earthquakes. The totality of the damaged features needs to be assessed.

Certainly, seismic shaking is the most important consideration in producing earthquake damage. The location of an ancient site relative to belts of seismic activity is important. Seismic geography or the current pattern of regional seismic activity is important for the assessment of the occurrence of past earthquakes as long as this does not limit ones thinking. It is erroneous to assume that, in the distant past, earthquakes could not have taken place in regions outside the active seismic belts of today.

5

## Earthquakes of Mexico

### 5.1     Pre-conquest Maya and Mexican codices

Native Mesoamerican manuscripts called "codices" ("codex" in the singular) have provided some information on the occurrence of natural phenomena, such as eclipses, volcanic eruptions, and earthquakes. The codices are conveniently divided into those that are pre-conquest or pre-Cortésian and those that are post-conquest. They can be further subdivided into those that treat solely the Maya and the Mexican realms. Extant pre-conquest Maya codices are limited to four and primarily deal with rituals, divinations, and the passage of time.

The known pre-conquest Maya codices are the *Dresden Codex*, the *Paris Codex*, the *Madrid Codex*, and the *Grolier Codex*. The Paris Codex is dated later than the Dresden Codex, and pictorially describes the ceremonies and rites of 11 successive katun cycles (7200 days each). It was presumably written sometime between the thirteenth and the fifteenth centuries. There is no mention of any earthquake occurrences. The Madrid Codex, written in the fifteenth century, is a ritualistic manual, similar to an almanac, conforming to the days of the year. Its subject matter includes world directions and other ritual subject matter but no specific reference to earthquakes. The Grolier Codex, dated by $^{14}$C measurements, was written in AD 1230 $\pm$ 130. The text contains tables pertaining to the planet Venus. Its pages show a standing figure facing a vertical column of Maya day signs. The earliest pre-conquest Maya codex known is the Dresden Codex with a date of compilation of about AD 1200–1250. This codex is also written in the form of an almanac. It contains multiplication tables for the synodical revolutions of Venus, tables for eclipses, depicts a variety of rituals and gods, and includes other tables of supposed divinatory or calendrical importance. It discusses many aspects of everyday Maya life such as disease and agriculture, and only indirectly (?) mentions earthquakes. The Maya calculated time with a *vigesimal* system calibrated by the synodical period (584 days) of Venus. There were 20 kin (days) in 1 uinal (month) and 18 uinals in 1 tun of 360 days (equivalent to 365 solar days minus 5 unfavorable days). The solar calendar of 365 days was broken into 18 months of 20 days followed by a final 5-day period, called *Uayeb*. In addition, the Maya used a ritual calendar of 260 days, the *tzolkin*. This ritual calendar was composed of 20 named days and 13 numbers and was intermeshed in a cogwheel fashion with the solar calendar. A particular combination of a numbered day with a month position could occur only every 52 years or 18 980 days and because of the permutations only 4 of the 13 named days out of 20 could ever be aligned with the first position of the 365-day solar calendar. A Maya date of 4 Ahau 3 Kankin refers to the numbered day in the ritual calendar followed by the month position in the solar calendar.

There are 20 tuns in 1 katun and 1 cycle represents 144 000 days or 20 katuns. A Maya Long Count was cumulative so a count of 9 cycles, 2 katuns, 10 tuns would be 1 314 000 days from the start of the Maya calendar, now known to be 3113 BC. It has been suggested that the worlds of the Maya end and are recreated at great cycles of 13 baktuns; 1 baktun is equivalent to 20 katuns, 400 tuns, or 144 000 days. Thirteen baktuns are equivalent to ~5128 years so the current cycle is believed to end at the Long Count of 13.0.0.0.0 4 Ahau 3 Kankin or December 21, 2012. In Maya mythology the end will come by flooding.

One of the 20 day signs of the ritual calendar was the *Caban*. Caban is the 17th day of the 20-day Maya month, corresponding to the Aztec day *Ollin*. *Ollin* is a Mexican day sign used to specify an earthquake. The sense of an earthquake is also found in the names of the 17th day in use elsewhere in Mesoamerica. In Quiché Maya, *noh* signifies great, tremendous, or earthquake. In Zapotec, *xoo* means powerful and *xòo*, *temblor de tierra*. Many of the other Maya day signs show a close similarity with the corresponding Aztec or Mexican day signs.

In the Maya language *cab* or *caban* signifies ground or earth but in the sense of volcanic area or earthquake region (De Gruyter, 1946; Seler, 1990). Landa in his treatise *Relación de las cosas de Yucatán* (1938), describes the caban sign as signifying *tierra* but also *ruido de los terremotos*. Joyce (1914, p. 249) states that *caban* seems to be the earthquake sign, used also to typify motion, as the Mexican *ollin*.

The *caban* sign, as a glyph, expresses a vertical direction as motion from above downward or from below upward. The Maya scribe illustrated the word *caban* by a curl hanging loosely over the cheek, conveying the idea of a thing shaking to and fro or a thing coming down, such as a tumbled down lock of hair. Some glyphic representations of the Mexican *ollin* sign and the Maya *caban* sign are shown in Figure 5.1 .

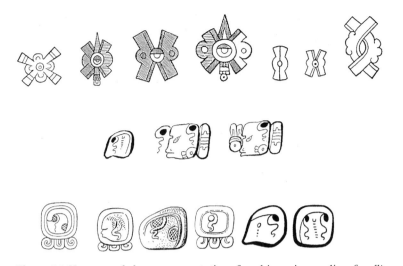

**Figure 5.1** Upper panel shows representations found in various codices for *ollin*, the 17th day sign in the Mexican ritual calendar, signifying earthquake or rolling movement (Seler, 1902). Lower panels show representations and usage as a hair curl on the Earth goddess of *caban*, the 17th Maya day sign found in various codices.

Figure 5.2 shows a page from the *Dresden Codex* with an image of the Maya God B seated in a house marked with numerous *caban* symbols. God B, known as a rain god, was a deity of life and creation. He was a universal deity to whom all natural phenomena were subject. The depiction of God B usually shows a long down-curling proboscis with the tongue or teeth hanging out in front. It is frequently depicted in association with the *caban* sign.

The term "pre-conquest Mexican", when referring to codices, refers to those regions that exclude the Maya regions. Pre-conquest Mexican codices, of which there are 11 known, are pictographic, ideographic, and of historical importance because they contain no European stylistic influence. Seven of these codices are from the Mixtec-speaking region, which was centered in the valley of Oaxaca in southern Mexico. The *Codex Vienna*, also known as the *Codex Vindobonensis Mexicanus I*, is a screenfold manuscript printed on deerhide which encompasses the period from the eighth century to the middle of the fourteenth century. It mainly covers the deeds and accomplishments of deities. The Mixtec had a number of important divinities, one of which was *4 Motion*. He is usually depicted (Figure 5.3) with round eyes, fangs in his open mouth, and pointing to the Mexican earth motion sign (*ollin*). Inasmuch as the original Mixtecs were believed to have come from the center of the Earth, this deity may be associated in some manner with the underworld of the Earth and probably symbolizes earthquakes.

Figure 5.2 Page 67 from the Dresden Codex showing utilization of the *caban* sign (Förstemann, 1880).

Monte Albán, near Oaxaca, was one of Mexico's earliest cities, and the seat of the Zapotec civilization. It started as an urban center around 500 BC, and had declined in importance for unknown reasons by AD 800, but was occupied when the Spaniards arrived in Oaxaca in AD 1521. The Zapotecs believed in forces, either natural or supernatural, that permeated their animistic religion. One god deserving of respect was *Pitao-Xòo*, or god of *temblors* of the earth. *Xòo* was also the 17th day name of the Zapotec version of the 260-day Mesoamerican ritual calendar.

It was accepted that the earth expressed its displeasure to the Zapotecs through Xòo by the rumbling of the surface and the opening of cracks and fissures in the ground. Xòo appears as an image on pottery and tombs at Monte Albán in a number of forms, either as a glyph for motion or as a mask having the snarling mouth of a feline with a cleft in its skull (Figure 5.4). The images and

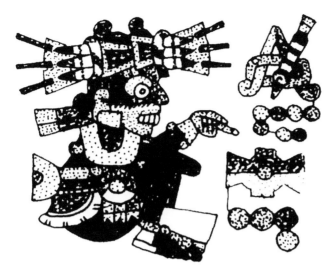

**Figure 5.3** The deity *4 Motion* as depicted in the Codex Vienna on page 33b. He is pointing to the *ollin* glyph signifying movement.

**Figure 5.4** Zapotec stylized masks found on ceramics and glyphs representing the 17th day, movement or earthquake. Note the similarity to the Mexican earth motion sign *ollin* (Caso, 1928; Marcus, 1992; Chan, 1993; Marcus and Flannery, 1996).

hieroglyphic captions accompanying carvings that depict Xòo do not appear to refer to places or specific earthquake occurrences, but rather to personal names or individuals associated by birth or occasion with the specific day of the ritual calendar. It may be that earthquake occurrences in the Valley of Oaxaca were of such frequency that individual events were not documented. This is a plausible explanation in view of the fact that the passage of time was not seen as linear, but rather cyclical and hence repetitive.

## 5.2    Post-Cortésian Mexican codices

No verified codices of the pre-conquest era exist for the Valley of Mexico itself, the heartland of Mesoamerica, with its familiar center Tenochtitlán (the site of the current capital Mexico City). It is indeed unfortunate that most pre-Cortésian codices were destroyed by overzealous Franciscan friars who did not want to remind the native people of their pre-Christian history and beliefs, which could inhibit their work of conversion. It is the early post-Cortésian Mexican codices which shed the most light on natural events such as earthquakes and volcanic eruptions of the fifteenth and sixteenth centuries. These codices were produced under Spanish patronage by native scribes, often contain year-by-year chronicles, and are excellent renditions of Aztec graphic art. Two examples containing superb pictorials are the *Codex Aubin* (*c.* 1576) and the *Codex Telleriano-Remensis* (*c.* 1562). The *Codex Ríos* contains pictorial descriptions of the same earthquake events illustrated in the Codex Telleriano-Remensis. This codex, which dates from 1566 to 1589, is also known as the *Codex Vaticanus A* or *Codex Vaticanus 3738*.

The *Codex Mendoza* (Berdan and Anawalt, 1997) was drafted in 1541, under the supervision of Spanish friars, and is a pictorial and textual account of early sixteenth-century Aztec life. References to earthquake occurrences can be found on folios 24v and 40r. Specific dates are not given but the codex covers the history of Tenochtitlán from AD 1325 to 1521.

Entries in these Mexican codices are laid out pictorially with the use of glyphs (*glifos* in Spanish; individual emblematic units) making use of an Aztec *xiuhpohualli* (year count) with the events arranged in sequential order, each one bearing the number and name of its particular year. The Aztec used a ritual calendar (*Tonalmatl*) of twenty 13-day periods or "weeks." These 260 days probably relate to the human gestation period. These 13 bundles of 20 days or 9 moons of ~29 days represented the ritual years. A year of 365 days was broken into 18 months of 20 days, each with 5 unfavorable days at the end. Each year, therefore, began with a so-called day sign 5 days later than the last; 365 is divisible by 13, with one as a remainder, so each year began with a day one in advance of the last. The lowest common multiple of 365 and 20 days is 5, which goes into 20 exactly 4 times, so the years began with one of four year signs only. The names of the years were *Tochtli* (*conejo*, rabbit), *Acatl* (*caña*, reed), *Tecpatl* (*pedernal*, flint), and *Calli* (*casa*, house), that is numbers 3, 8, 13, and 18 in the set of 20 day names. These year names would repeat themselves in cycles of 4 over 52 years (4 × 13). Hence, the cycle that begins as 1 Tecpatl, 2 Calli, 3 Tochtli, 4 Acatl, 5 Tecpatl, etc., ending with 13 Acatl, starts again with 1 Tecpatl after 52 years have elapsed. Year 13 Acatl corresponds to the creation of the "Fifth Sun" in the year AD 1479 during the reign of Ayayacatl and 1 Acatl correlates with AD 1519, the date of Cortés' arrival. This linear chronology is valid for events of the 1400s and 1500s

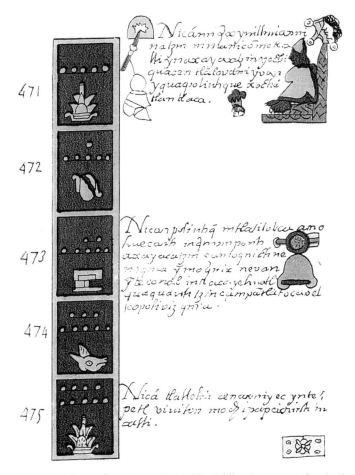

**Figure 5.5** A page from the Codex Aubin (Códice de 1576) stating in the bottom panel that an earthquake took place in the year 9 Acatl (reproduced by permission from Historia de la Nacion Mexicana, 1963).

because of the independent corroboration of many of the described events. Extrapolation to earlier dates is artificial and sometimes leads to debate about the exact times and years in the genealogy of tribal ancestors.

For the depiction of an earthquake occurrence, use was made of the *ollin* and *tlalli* glyphs (*glifo de la tierra* = *tlalli*, a rectangle representing a patch of land *el glifo de movimiento* = *ollin*). The date of occurrence was shown by a number count and the pictograph for *Tochtli, Acatl, Tecpatl,* or *Calli.* Occurrence in the day or night was indicated by filling in or leaving empty the center of the *ollin* glyph thereby representing a central *Glifo de tonatiuh* (the sun). The glyphs were briefly accompanied by a written annotation in Nahuatl or Spanish. Some examples from the Codices Aubin and Telleriano-Remensis are shown in Figures 5.5 and 5.6.

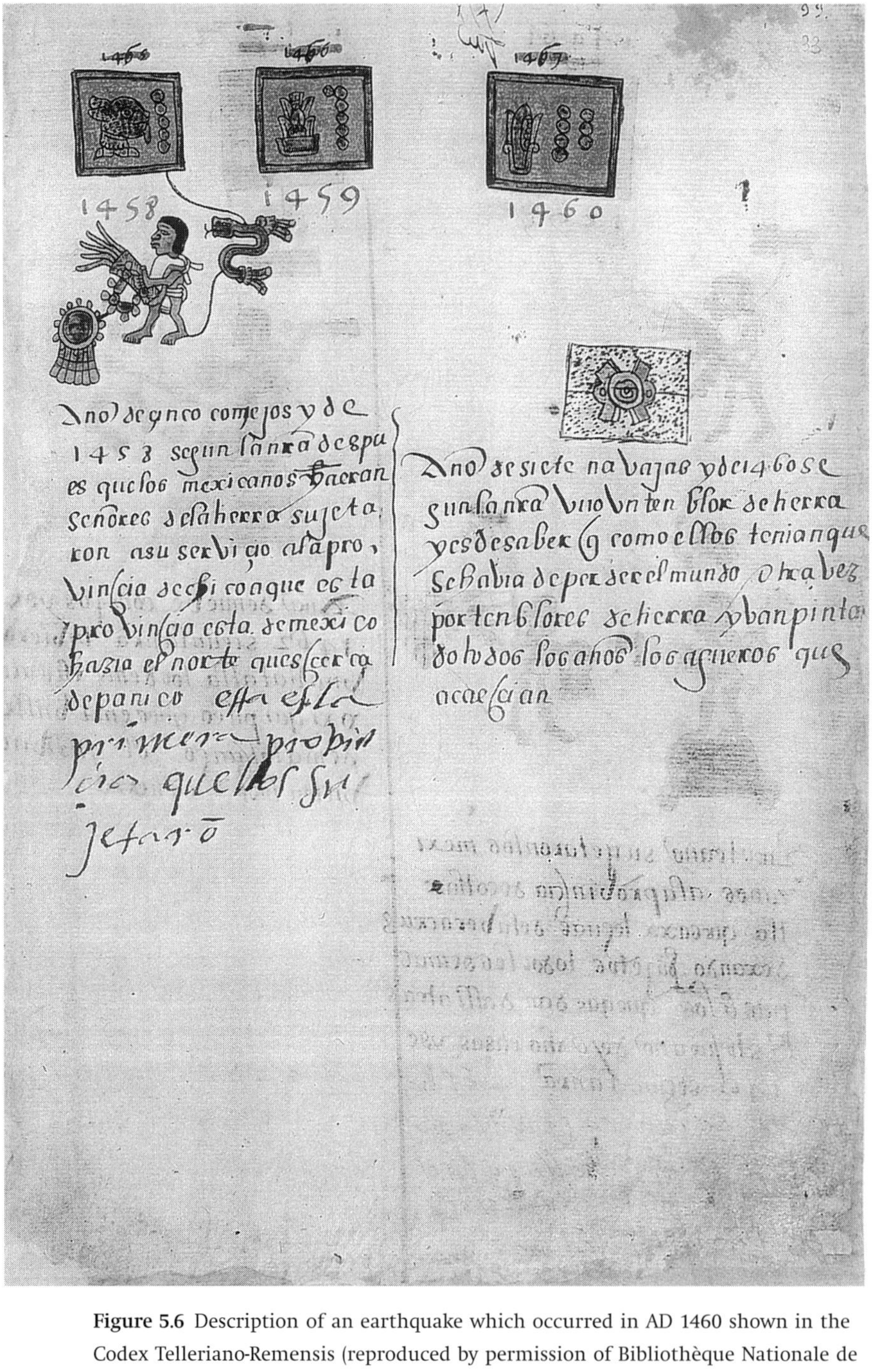

**Figure 5.6** Description of an earthquake which occurred in AD 1460 shown in the Codex Telleriano-Remensis (reproduced by permission of Bibliothèque Nationale de France). "In the year of 7 Flint and 1460 an earthquake occurred. It deserves to be remarked that since, according to their belief the world was again to be destroyed by earthquakes, they recorded in their paintings each year, the omens that occurred."

The bottom panel in Figure 5.5 from the Codex Aubin depicts an earthquake that took place in the year 9 Acatl (AD 1475) and is annotated in Nahuatl: "Nicā tlallolin cenca miyec yn tepetl viuiton mochi papachiuh in calli." This translated as "Here [the Earth] shook much. Many hills were crumbled and all the houses were crushed." The page from the Codex Telleriano-Remensis shown in Figure 5.6 illustrates an earthquake that took place in the year 7 Tecpatl (AD 1460); it is annotated in Spanish that since it was believed that the earth would be destroyed again by earthquakes, every year they painted the omens that occurred: "Año de siete navajas y de 1460 según la (cuenta) nuestra, hubo un temblor de tierra y es de saber que como ellos temían que se había de perder el mundo otra vez por temblores de tierra, iban pintando todos los años los agüeros que acaecían." This text is translated in full in Figure 5.6 legend.

In addition to earthquake occurrences, glyphs were used to depict solar eclipses, comets, columns of smoke, snowstorms, and windstorms. Figure 5.7 shows a folio from the Codex Telleriano-Remensis covering the years 1507–1509. Several historical and geophysical events are shown and extensively annotated on this page. In the year 2 Acatl (1507) the images show an earthquake (*ollin*) and the occurrence of a partial solar eclipse (the glyph of the sun connected by a line to the *tllali* glyph). The year 1507 also depicts and describes the drowning of 2000 soldiers (each tree representing a count of 500) in the Tuçac (Tozac) River (a river flowing between the Mexican States of Puebla and Oaxaca) as they were en route to Mixteca (in the northwest part of Oaxaca) to conquer some provinces. Year 4 Calli (1509) shows the occurrence of a celestial phenomenon. The words describe a *mexpanitli* or cloud banner (cloud of smoke?) as a brilliant light that was seen in the eastern sky for over 40 days. A later annotation adds that the cloud banner preceded the return of *Quetzalcoatl* whom the Aztecs seem to have associated with the forthcoming arrival of Cortés. The year of 6 Calli (1537) also deserved some historical documentation (Figure 5.8). An earthquake occurrence and smoke curls shown rising above a star are depicted, possibly smoke from Popocatepetl (Nahuatl for smoking hill): "Este año de seis casas y de 1537 se quisieron alzar los negros en la ciudad de México . . . Humeaba la estrella y hubo un temblor de tierra, el mayor que yo he visto, aunque he visto muchos por estas partes." This translates as "This year of six houses [6 House] and 1537 the blacks tried to rebel, in the city of Mexico the instigators were hanged. The star was smoking and there was an earthquake, the worst I have seen, even though I saw many of them in these lands."

Table 5.1 gives a compilation of pre-Cortésian Mexican earthquakes begin-ning in the year 1455. It is of course not possible to know the size of these pre-Cortésian earthquakes but it should be safe to assume that they were of sig-nificant size to have deserved historical mention. There were 19 events for the

**Figure 5.7** Folio 42r of the Codex Telleriano-Remensis depicting historical and geophysical events that took place in the years 1507–1509 (reproduced by permission of Bibliothèque Nationale de France).

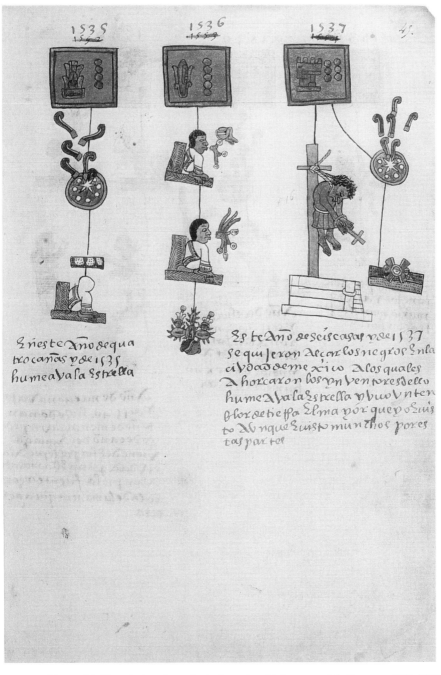

**Figure 5.8** Historical and geophysical events which occurred in the years 1535–1537 as depicted in the Codex Telleriano-Remensis (reproduced by permission of Bibliothèque Nationale de France).

Table 5.1. *Pre-Cortésian Mexican earthquakes*

| Year | Description |
| --- | --- |
| 1455 | Valley of Mexico ". . . hubo también terremoto . . ." (*Anales de Tlatelolco*, 1948). |
| 1460 | Valley of Mexico (*Códice Telleriano-Remensis*, 1995). |
| 1462 | Valley of Mexico ". . . hubo un temblor de tierra." (*Códice Telleriano-Remensis*, 1995). |
| 1468 | Valley of Mexico (*Códice Telleriano-Remensis*, 1995). |
| 1469 | Coast of Anáhuac, province of Xochitepec [State of Morelos] (Torquemada, 1969). |
| 1474 | Imperial Mexico (Orozco y Berra, 1887). |
| 1475 | "En el año 9 acatl, 1475, un fuerte temblor . . ." (*Codex Mexicanus*, 1939; *Códice Aubin*, 1963). |
| 1476 | Imperial Mexico (Acosta and Reynoso, 1996). |
| 1480 | "Fuerte temblor generalizado." (Acosta and Reynoso, 1996; *Códice Telleriano-Remensis*, 1995). |
| 1487 | "En el año 1487, en que un gran temblor de tierra . . ." (Acosta and Reynoso, 1996). |
| 1489 | "Fuerte temblor en todo el Imperio." (*Codex Mexicanus 1939*; *Códice Aubin*, 1963). |
| 1490 | "Al cuarto año del reinado de Ahuizotl dicen que tembló . . ." (Torquemada, 1969). |
| 1495 | Imperial Mexico (*Códice Telleriano-Remensis*, 1995). |
| 1496 | "En 4 tecpatl hubo un temblor general llenándose la tierra de grietas." (*Códice Aubin*, 1963). |
| 1499 | Valley of Mexico. "And on that day the earth shook four times." (Bierhorst, 1992a). |
| 1507 | "Año de los cañas y de 1507, hubo un eclipse de sol y tembló la tierra . . ." (*Códice Telleriano-Remensis*, 1995). |
| 1510 | Oaxaca, Mexico (Acosta and Reynoso, 1996). |
| 1511 | "En este año hubo grandes nieves y tembló la tierra tres veces." (*Códice Telleriano-Remensis*, 1995). |
| 1513 | Valley of Mexico (*Códice Telleriano-Remensis*, 1995). |
| 1517 | Volcanic earthquake. "El Popo, precedido por movimientos tectónicos, hace erupción." (Sugawara, 1987). |
| 1519 | Volcanic earthquake. "El Popo en actividad provoca microseismos." (Sugawara, 1987). |
| 1523 (April 1) | "El primero terremoto de que se ve hecha mención por los españoles immediatamente después de la conquista . . ." (Acosta and Raynoso, 1996). |

58-year interval from 1455 to 1513, giving an average of one event every 3 years or so.

## 5.3   Earthquake recurrence in the modern era

The earthquakes of Mexico fall into two categories. One group contains the subduction-zone earthquakes of shallow (15 to 20 km) and intermediate (65 to 150 km) focal depths that occur along the Jalisco, Michoacan, Guerrero, and Oaxaca segments of the Pacific coast. The second group comprises shallow

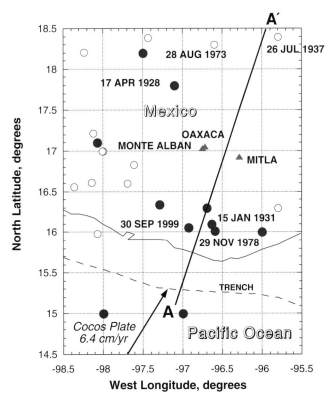

**Figure 5.9** Location map showing significant epicenters and magnitudes of earthquakes in the Oaxaca area, 1900–2000. Solid circles, 7.5 ≥ M ≤ 8.1; open circles, 7.0 ≥ M ≤ 7.4. Data are given in Table 5.2. Line A–A′ shows location of cross-section shown in Figure 5.10.

crustal events at a latitude of 19° to 20° N along the west-to-east Trans-Mexican volcanic belt. It is the large subduction-zone earthquakes that have been responsible for most of the damage to Mexican cities, communities, and archaeological sites.

The Oaxaca area experiences along its coast about one-third of the total seismic activity of Mexico. Earthquakes here are the result of the subduction of the Cocos plate beneath the North American plate. Figure 5.9 shows the location of the large-magnitude shocks (M ≥ 7.0) that have occurred since 1903 in the vicinity of Oaxaca. Table 5.2 lists the earthquake epicenters and magnitudes of the shocks in Figure 5.9. The open and solid circles in Figure 5.9 represent earthquakes with magnitudes of 7 to 7.4 and of 7.5 to 8.1, respectively. The arrow shows the direction of relative plate motion at a rate of 6.4 cm per year. A section A–A′ across the Oaxaca subduction zone is shown in Figure 5.10. The dip of the subducting plate is approximately 15° in a north-northeast direction resulting

Table 5.2. *Large earthquakes (M ≥ 7.0) of Mexico (1900–2000) in the proximity of Oaxaca, only events between 15° to 19° N and 95.5° to 98.5° W are listed*

| Date | Latitude (°N) | Longitude (°W) | Magnitude |
| --- | --- | --- | --- |
| January 14, 1903 | 15.00 | 98.00 | 8.1 |
| December 29, 1917 | 15.00 | 97.00 | 7.7 |
| January 4, 1920 | 18.20 | 97.50 | 7.8 |
| March 22, 1928 | 16.00 | 96.00 | 7.5 |
| April 17, 1928 | 17.80 | 97.10 | 7.7 |
| June 17, 1928 | 16.30 | 96.70 | 7.9 |
| August 4, 1928 | 16.83 | 97.61 | 7.4 |
| October 9, 1928 | 16.34 | 97.29 | 7.6 |
| January 15, 1931 | 16.10 | 96.64 | 7.8 |
| July 26, 1937 | 18.40 | 95.80 | 7.3 |
| December 23, 1937 | 17.10 | 98.07 | 7.5 |
| January 6, 1948 | 17.00 | 98.00 | 7.0 |
| January 6, 1948 | 17.00 | 98.00 | 7.0 |
| December 14, 1950 | 17.22 | 98.12 | 7.3 |
| August 23, 1965 | 16.30 | 95.80 | 7.3 |
| August 2, 1968 | 16.60 | 97.70 | 7.1 |
| August 28, 1973 | 18.30 | 96.60 | 7.2 |
| November 29, 1978 | 16.01 | 96.59 | 7.7 |
| October 24, 1980 | 18.21 | 98.24 | 7.0 |
| June 7, 1982 | 16.61 | 98.15 | 7.2 |
| June 7, 1982 | 16.56 | 98.36 | 7.0 |
| February 25, 1996 | 15.98 | 98.07 | 7.1 |
| June 15, 1999 | 18.39 | 97.44 | 7.0 |
| September 30, 1999 | 16.06 | 96.93 | 7.5 |

Data compiled from Singh *et al.*, (1981, 2000) and National Earthquake Center.

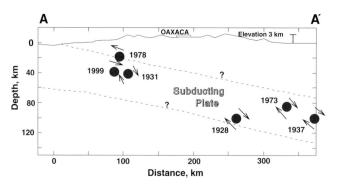

**Figure 5.10** A section along the line A–A' shown in Figure 5.9. Except for topography, there is no vertical exaggeration. Hypocenters occur within the subducting plate and on top of the plate boundary near the coastline.

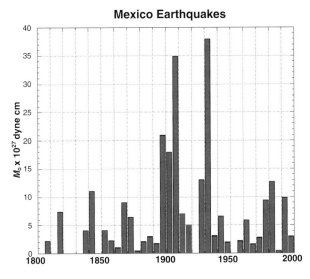

**Figure 5.11** Seismic moment release as a function of time from 1800 to 2000 for the earthquakes of Mexico.

in deeper earthquake hypocenters further inland. Events that occur within the downgoing slab are the result of downdip tension whereas the 1978 event was the result of reverse slippage on the upper interface of the subducting plate.

Mexico has a fairly well-documented recorded history of great earthquakes (those with magnitudes of 7 or greater) from 1800 to 1996. The inferred or determined magnitudes can be converted to seismic moment, a more appropriate measure of earthquake size, using the relation: $\log M_0 = 1.5M + 16.1$. A plot of the seismic moment release versus time for earthquakes within the Mexican subduction zone is shown in Figure 5.11. It can be seen that the release of seismic moment as a function of time took place in bursts. This observation leads to the suggestion that the occurrence of large earthquakes in Mexico, along the subduction zone, is clustered in time. If temporal clustering does indeed exist over long periods of time our ability to estimate the probability of occurrence of future earthquakes within a specified time interval is increased. From about 1800 to 1945 the seismic moment release took place in bursts of about 15 years of activity followed by relative quiescence or lesser seismic moment release for the next 15 years. This predictive trend suggests that subsequent clustering would be centered about 1966 or so but this was not observed. The clustering of seismic moment release in the early 1980s lies outside this window. One explanation is that the high concentration of moment release around 1910 and 1930 may have produced a large reduction in stress buildup along this segment of the convergent

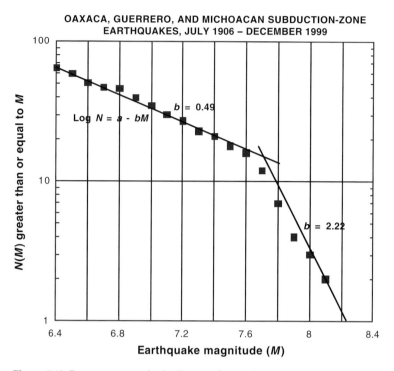

**Figure 5.12** Frequency–magnitude diagram for Mexican subduction-zone earthquakes. N is the cumulative number of earthquakes greater than or equal to a given magnitude; M is surface-wave magnitude or moment magnitude; a is the number of earthquakes greater than or equal to M = 0 and b is the slope of the lines shown.

plate boundary along the Mexican coast and that sufficient time had not elapsed for subsequent accumulation of stress prior to the next major burst of activity.

The frequency–magnitude statistics (Figure 5.12) of Mexican subduction-zone earthquakes of magnitude greater than or equal to 6.4 are believed to be complete for the 92.5-year interval from July 1906 to December 1999. With this caveat one can make some inferences concerning the return time of large earthquakes in this region. The key observation is that the familiar frequency–magnitude relation, $\log N = a - bM$, shows a break in slope at a magnitude of 7.7, from a slope of $b = 0.49$ below this magnitude, to a slope of $b = 2.22$ above. This signature means that earthquakes above and below this magnitude value belong to different populations. As a consequence, earthquakes of magnitude 7.7 and greater occur less frequently than would be indicated by extrapolation using the numbers of smaller events. In addition, an $M = 7.7$ event can be taken as the most likely maximum size event to strike this segment of the subduction plate boundary. Based on these data, the expected return time for an event of

$M = 7.7$ or greater is about 6 years or so, somewhat less frequent than is sug-gested from the data of the pre-colonial period. An $M = 8$ event would have an expected average return time of 31 years. There is no reason, however, to expect the rates of occurrences from these two temporal samples of Mexican earthquakes to be in any closer agreement unless the data were stationary in time over several centuries and all of the pre-colonial events were of the same corresponding magnitude.

The occurrence on a time scale of decades, of large-magnitude earthquakes that originate along the coastal region of Mexico, causing damage to cities in the interior, suggests strongly that many of Mexico's archaeological sites must also contain evidence for past earthquake occurrences. This evidence may have been overlooked or as yet may be undiscovered. The argument is strengthened further by the fact that recent earthquakes such as the $M_w = 7.5$ Mexico earthquake of September 30, 1999 produced damage to pyramid walls at the Monte Albán archaeological site near Oaxaca, Mexico.

# 6

## Earthquakes in the Maya empire

### 6.1    Geographic and tectonic setting

The Maya empire encompasses the southeastern portion of Mexico in-cluding all of the Yucatán peninsula, Guatemala, Belize, the western parts of El Salvador, and Honduras. Topographically the Maya were positioned in three environmental zones, the Pacific coastal plain and hilly piedmont region, the highlands, and the lowlands. The northern highlands are composed primarily of metamorphic rocks, whereas the southern highlands are a zone of spectacu-lar and recent volcanic topography. The southern highlands experience frequent seismic and volcanic activity. Young volcanoes form the spine or continental di-vide that runs parallel to the Pacific coast from Mexico southeasterly through Guatemala, El Salvador, Nicaragua, and Costa Rica (Figure 6.1). North of the belt of volcanic activity is the rugged older volcanic highland region possessing valleys and basins filled with fertile volcanic soils. The lowlands are subdivided into the southern, central, and northern regions. All of the Yucatán peninsula forms the northern lowlands.

Running southeasterly through Mexico, roughly parallel to the Middle America subduction zone, are a number of seismically active zones. These zones intersect the landward extension of the currently active Caribbean seismic plate boundary that passes westward through Guatemala. Many of the sites occupied by the Maya would have been subjected to the effects of earthquakes originating in these zones (Figure 6.2).

Earthquakes that take place in the regions occupied by the Maya can be placed in three categories. The first category includes the large subduction zone earth-quakes ($M < 8.5$), which are generated as a result of the north-eastward-moving Cocos plate being thrust under the Caribbean and North American plates. This

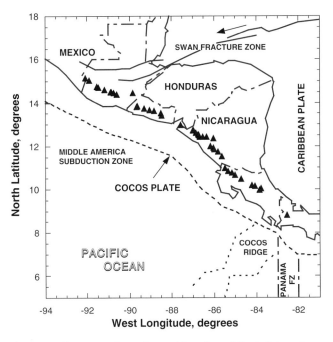

**Figure 6.1** Plate tectonic setting and location of Central American volcanoes.

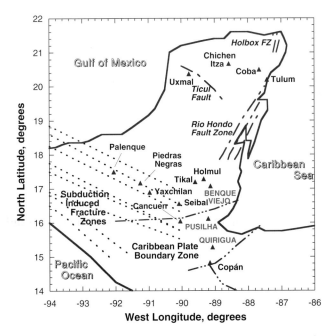

**Figure 6.2** Tectonic setting of the Maya area and the location of some archaeological sites.

subduction begins offshore at the Middle America Trench which parallels the Pacific coast of Central America. Earthquake hypocenters that occur along the subducting plate are shallow-focus offshore, increasing to depths of several hundred kilometers beneath the coastline. These earthquakes have epicentral distances to population centers ranging from 80 to 100 km, and typically have magnitudes of 7 or greater. The second class of earthquake is the more frequently occurring shallow-focus (5 to 15 km in depth) earthquake affiliated with the inland volcanic zone that parallels the Pacific coast. These earthquakes range in magnitude from 5.7 to 6.5, and are usually destructive because of the proximity of their hypocenters to major population centers. Damage is severe because the earthquake ground accelerations are amplified by the surficial deposits of volcanic ash found along the volcanic front. These earthquakes repeatedly cause damage to cities, towns, and communities that are situated along the volcanic alignment.

Volcanic eruptions are at times accompanied by felt earthquakes, but there is no evidence of damaging earthquakes correlating with eruptive activity. Destructive earthquakes originating along the volcanic front appear to be the result of applied regional tectonic stresses rather than the stresses produced by the subterranean movement of magma. The final locus of earthquake activity is related to the left-lateral relative motion between the Caribbean and North American plates. These earthquakes are shallow-focus and can be quite damaging with magnitudes of 7 or greater and repeat or recurrence times of 100 to 200 years.

## 6.2    Historical seismic and volcanic activity

Volcanic eruptions and earthquakes have long plagued the countries of Guatemala and El Salvador. Prior to the arrival in 1524 of Pedro Alvarado, a captain of Hernando Cortés, documentation of volcanic eruptions and earthquakes is scant. Some descriptions of early tectonic disturbances can be found in Rockstroh (1883), Montessus de Ballore (1885), Sapper (1925), and Wells (1857).

The capital of Guatemala was destroyed three times, each time after being located in a different spot. On September 10, 1541, earthquakes, and an accompanying rain-induced lahar from the side of the volcano Agua, destroyed the first capital "Santiago de los Caballeros de Guatemala." The Cakchiquel, a Maya tribe of the highlands that entered Guatemala around AD 1100 from Tabasco, Mexico, recorded the event in their annals (Recinos *et al.*, 1953) with some glee: "On the day 2 knife the volcano Hunahpú [Agua] swept down . . . the water gushed . . . the Castilians perished and died . . ." The second capital, today known as Antigua, was positioned in the same valley somewhat further away from the base of the

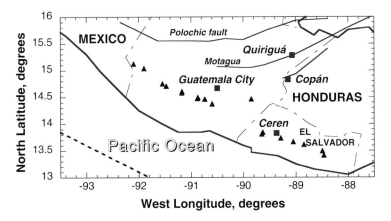

**Figure 6.3** Location of the Maya site of Ceren that was buried by a regional ash fall around AD 600. Triangles are locations of active volcanoes.

volcano. Antigua grew in the middle of the eighteenth century to about 60 000 residents and was a center for ecclesiastical activities of the Catholic Church, with many convents and churches. Antigua was subjected to a long sequence of damaging earthquakes culminating in four disastrous events in the latter part of 1773. An *earthquake psychosis* permeated the community and the religious orders did not look favorably on the secular decision to move the capital of Antigua and its power base to a new location in 1776, some 32 km to the east in a neighboring valley. The rest is history. In spite of repeated earthquake activity, Guatemala City flourished under the false impression of impunity, only to be destroyed in the earthquakes of 1917–1918, which left nearly 100 000 people homeless.

As can be seen, the Maya not only occupied varying topographic regions but also regions of different seismic and volcanic geography. Inhabitants of the coastal plain and the highlands must always have been subjected to the effects of earthquakes and volcanic eruptions. A particular case in point is Ceren, a Maya site about 30 km west of Lake Ilopango in El Salvador (Figure 6.3). Ceren was occupied around AD 590, near the end of the Maya Early Classic Period (AD 250–600). It was built on the volcanic ash deposited by an earlier massive eruption of Ilopango around AD 175. Ceren was subsequently buried by a small, regional volcanic ash fall around AD 600 resulting in a site of outstanding archaeological preservation. What was remarkable about the Ceren site was the accidental discovery of well-preserved thatched roofs, which had collapsed to the floor from the weight of volcanic ash (Sheets, 1971a, b, 1992). Earthquakes are a common occurrence in El Salvador in the vicinity of the Ceren site, but no evidence for earthquake activity concurrent with the volcanic eruption could be found. Round-bottomed ceramic serving bowls were found undamaged *in situ* on the tops of flat walls and shelves, suggesting sudden burial. The residents of Ceren also used a type of construction, *bajareque*, that is very earthquake resistant. The bajareque construction is a mud-filled wooden lattice wall or timber skeleton erected on a flat clay platform; this framework is then reinforced from within with tough but flexible vertical rods, which are horizontally reinforced with other sticks lashed to the rods. Such structures stand up well during earthquake shaking. The implications of the volcanic disaster that struck Ceren are still being debated. A large regional ash fall would have undoubtedly produced agricultural deprivation and increased the susceptibility to subsequent flooding because of the damage to binding vegetation. Increased rainfall is a common occurrence after large eruptions of volcanic ash. Particles that are thrown into the air from volcanoes in tropical zones serve as the nuclei for the formation of water droplets. Archaeologists and anthropologists argue that the volcanic disaster produced migration or the displacement of a large number of Maya people to other parts of Central America. Such a migration would have had an impact on the social fabric of other communities.

## 6.3    Earthquakes at Quiriguá, Guatemala

Evidence of past earthquake occurrences can be found at Quiriguá, in eastern Guatemala. Quiriguá was a Maya site which controlled a lucrative trade route along the Motagua River Valley. Jade and obsidian were transported from the highlands to the Caribbean coast. Early construction began in AD 550 during the Terminal Classic Period with Quiriguá reaching eminence between AD 724 and 784. It reached its final era about AD 810 or 9.19.0.0.0 in the Maya Long Count.

The San Agustin, Motagua, and Polochic fault zones (Figure 6.4) represent the western landward extension of the plate boundary between the North American and the Caribbean lithospheric plates. However, there is a relative paucity of major historical earthquakes associated with this zone. Most of the damaging earthquakes of Guatemala have taken place in the coastal plain and highland region paralleling the Pacific coast. A destructive event took place in 1765 with a presumed epicenter somewhere near a longitude of 89.5° W. The earthquake of July 29, 1773, centered near the western end of the Motagua fault, destroyed the colonial capital of Antigua, Guatemala. Severe damage was reported from the 1856 Omoa, Honduras event, which may well be the largest historical earthquake ever to have occurred in this vicinity. Minor damaging shocks took place in 1929 near Puerto Barrios and in 1945 near Quiriguá where ground cracking and damage to the United Fruit Company hospital were reported (Figure 6.5).

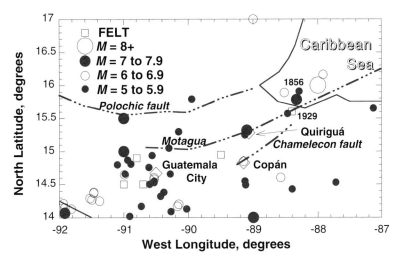

**Figure 6.4** Historical seismicity of Guatemala showing location of the $M = 7.5$ earthquake of 1976 near the famous Maya site, Quiriguá. The San Agustin fault zone lies in very close proximity to the Motagua fault zone. The loci of previous damaging Guatemalan earthquakes are also shown. Data are given in Table 6.1.

At 3:02 a.m. on February 4, 1976, a devastating $M = 7.5$ earthquake occurred with an epicentral location near Quiriguá. This earthquake produced an average left-slip offset of 1.1 m. There were 24 000 fatalities and 76 000 people were injured in the highlands of Guatemala. Much of the damage to life and property in the vicinity of Quiriguá took place in the elevated areas of the Motagua River drainage basins southwest of the ruins. The reasons were primarily architectural. Much of the construction was adobe with heavy tile roofs. Structures in the lower Motagua valley fared better because of the wood framing and roofs that were made of thatch or sheet metal. Ground rupture was documented near the Quiriguá ruins (Figure 6.6).

Seismological analyses of the $M = 7.5$ Guatemala event determined a seismic moment ranging from $2 \times 10^{27}$ to $2.6 \times 10^{27}$ dyne cm. One can easily estimate the average recurrence rate for events of this size along this segment of the Caribbean plate boundary. Global positioning measurements reveal that the Caribbean plate is moving eastward relative to North America at a rate of 21 $\pm$ 1 mm per year (Dixon *et al.*, 1998). Over the past 2000 years if all of the accumulated slip along the plate boundary were expended seismically over an assumed rupture zone of 400 km in length and 15 km in width, a cumulative seismic moment of $7.6 \times 10^{28}$ dyne cm would have been released. What does the historical record show? Our knowledge is incomplete but Table 6.1 lists the major earthquakes known to have occurred within the Caribbean plate boundary zone through Guatemala from 1765 to 2001.

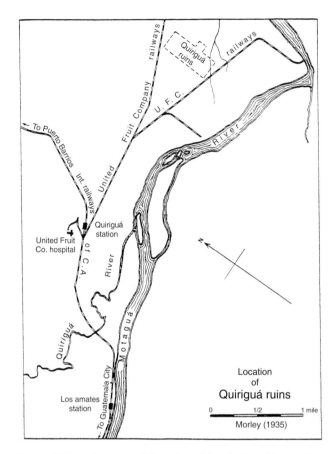

**Figure 6.5** Location map of the ruins of Quiriguá and its relation to the Quiriguá station and the Motagua River. From Morley (1935) courtesy of the Carnegie Institution of Washington.

Converting the values of $M$ to $\log M_0$ in Table 6.1 using the scaling law $M = M_s = M_w = (2/3) \log M_0 - 10.73$, reveals that the amount of moment release from 1765 to 2001 was $3.0 \times 10^{28}$ dyne cm of the total moment release predicted for this segment of the Caribbean plate boundary over the past 2000 years. A slip deficit of $4.6 \times 10^{28}$ dyne cm translates to the equivalent of twenty $M = 7.5$ events or five $M = 7.9$ events having occurred in the earlier historical part of this period, or a substantial amount of slip having been released as a result of slow aseismic sliding.

It is difficult to believe that, with such a recurrence time, earthquake effects were not produced at Maya sites in proximity to the Caribbean plate boundary. It is most likely that earthquakes were a continuing problem for the seventh- to ninth-century Maya populations at Quiriguá and other nearby sites such as

**Figure 1.1** A portion of the Seismographic Map of the World prepared by Mallet and Mallet in 1857.

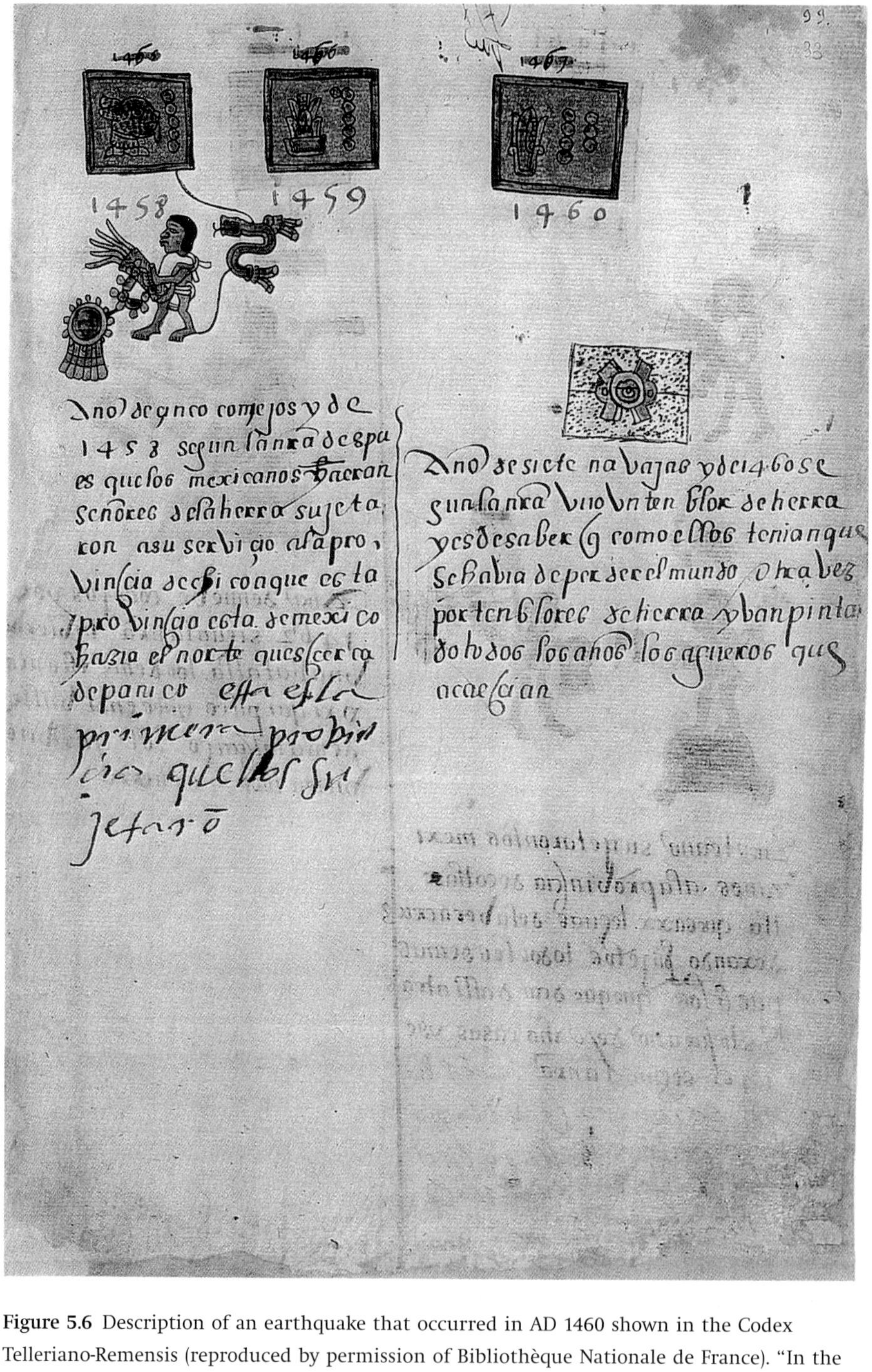

**Figure 5.6** Description of an earthquake that occurred in AD 1460 shown in the Codex Telleriano-Remensis (reproduced by permission of Bibliothèque Nationale de France). "In the year of 7 Flint and 1460 an earthquake occurred. It deserves to be remarked that since, according to their belief the world was again to be destroyed by earthquakes, they recorded in their paintings each year, the omens that occurred."

**Figure 5.7** Folio 42r of the Codex Telleriano-Remensis depicting historical and geophysical events that took place in the years 1507–1509 (reproduced by permission of Bibliothèque Nationale de France).

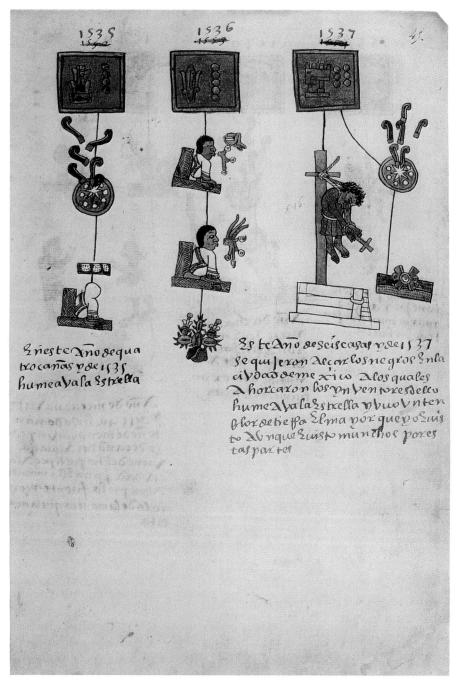

**1535**

**1536**

**1537**

Ɛ nes te Xno de qua
tro cañas y de 1535
hu mea ya fa Ɛs tre fa

Ɛs te Ano de seis casas y de 1537
se qui Jeron Alcar los negros Ɛn la
ciydad de me xico Alos quales
A horcaron los yn ven to res de lco
hume Abala Ɛs trella y buo Vn ten
blor de tie ffa Ɛl ma yor que yo Ɛuis
to Aunque Ɛuisto munchos pores
tas partes

**Figure 5.8** Historical and geophysical events that occurred in the years 1535–1537 as depicted in the Codex Telleriano-Remensis (reproduced by permission of Bibliothèque Nationale de France).

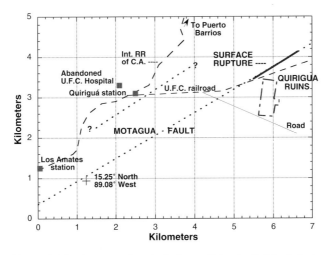

**Figure 6.6** During an archaeological expedition to Quiriguá during 1976, *en echelon* ground rupture of about 1 km in length with an offset of 6 cm was measured close to the ruins at Quiriguá. A railroad bed disrupted by about a few centimeters was noted near the Quiriguá station (Bevan and Sharer, 1983). UFC, United Fruit Company; Int. RR of CA, International railroad of California.

Copán. Let us look at the evidence for past earthquake damage at Quiriguá, beginning with a discussion of the earthquake damage produced by the 1976 Guatemala event.

The site at Quiriguá is famous for its unusual and large carved zoomorphs and the height of its sandstone stelae which were erected to record religious and historic events. It is particularly notable for an immense Stela E (Monument 5, M5, in Figure 6.7) which rises 8 m above ground with its butt extending 2.4 m below ground. It was erected in AD 771 and is the largest block of stone ever carved by the Maya. Figure 6.7 shows the layout of the site at Quiriguá together with the locations of some noted earthquake effects. The famous monuments are located in the Great Plaza north of the Acropolis and Ballcourt Plaza. Bevan and Sharer (1983) documented the damage to some monuments and structures that took place during the 1976 earthquake. The earthquake collapsed some archaeological excavations and continuing aftershocks restricted activity for about a month afterward. The structures damaged were masonry structures in the Acropolis and two monuments in the Great Plaza. Partial masonry collapse of both the exterior and interior walls took place in Structure 1B-1, the most southerly building in the Acropolis (Figure 6.8). Bevan and Sharer also pointed out that the roof vaults and upper vault of this structure and others nearby had probably collapsed from the shaking of earlier earthquakes. Masonry damage was also noted in Structures 1B-4 and 1B-5.

Table 6.1. *Earthquakes in the vicinity of the Caribbean plate boundary zone, 1765–2001*

| Date | Latitude (° N) | Longitude (° W) | Magnitude | Reference |
|------|------|------|------|------|
| 1765 | 14.95 | 89.50 | Destructive | NEIC[a] |
| July 29, 1773 | 14.90 | 90.80 | Felt | NEIC |
| July 22, 1816 | 15.50 | 91.00 | 7.6 | White (1985) |
| August 4, 1856 | 16.00 | 88.00 | 8.0 | NEIC, Sutch (1981) |
| September 3, 1874 | 14.50 | 90.70 | Felt | NEIC |
| June 12, 1912 | 17.00 | 89.00 | 6.8 | Gutenberg & Richter (1949) |
| March 8, 1913 | 15.00 | 91.00 | Felt | NEIC |
| September 7, 1915 | 14.00 | 89.00 | 7.9 | NEIC |
| December 29, 1915 | 14.60 | 88.58 | 6.4 | White & Harlow (1993) |
| December 26, 1917 | 14.53 | 90.53 | 5.8 | White & Harlow (1993) |
| December 29, 1917 | 14.55 | 90.53 | 5.8 | White & Harlow (1993) |
| January 3, 1918 | 14.50 | 90.60 | 5.8 | White & Harlow (1993) |
| January 4, 1918 | 14.58 | 90.53 | 6.1 | White & Harlow (1993) |
| January 25, 1918 | 14.60 | 90.53 | 6.2 | White & Harlow (1993) |
| April 28, 1919 | 14.50 | 91.00 | Felt | NEIC |
| February 4, 1921 | 15.00 | 91.00 | 7.5 | NEIC |
| July 7, 1930 | 14.17 | 90.17 | 6.0 | White & Harlow (1993) |
| July 14, 1930 | 14.20 | 90.15 | 6.9 | White & Harlow (1993) |
| July 17, 1930 | 14.13 | 90.03 | 5.7 | White & Harlow (1993) |
| December 3, 1934 | 14.85 | 89.15 | 6.2 | White & Harlow (1993) |
| August 10, 1945 | 15.25 | 89.13 | 5.7 | White & Harlow (1993) |
| December 31, 1974 | 14.15 | 91.92 | 6.0 | NEIC |
| December 31, 1974 | 14.13 | 91.82 | 6.1 | NEIC |
| February 4, 1976 | 15.32 | 89.10 | 7.5 | NEIC |
| February 4, 1976 | 14.94 | 90.56 | 5.4 | NEIC |
| February 4, 1976 | 14.17 | 90.73 | 5.0 | NEIC |
| February 6, 1976 | 14.32 | 90.43 | 5.0 | NEIC |
| February 6, 1976 | 14.76 | 90.61 | 5.7 | NEIC |
| February 8, 1976 | 15.57 | 88.47 | 5.6 | NEIC |
| February 9, 1976 | 15.32 | 89.07 | 5.2 | NEIC |
| March 7, 1976 | 14.81 | 90.89 | 5.1 | NEIC |
| March 9, 1976 | 14.87 | 90.94 | 5.2 | NEIC |
| March 13, 1976 | 14.80 | 91.10 | 5.1 | NEIC |
| February 22, 1978 | 14.25 | 91.38 | 6.4 | NEIC |
| March 30, 1978 | 15.05 | 90.30 | 5.1 | NEIC |
| July 29, 1978 | 14.66 | 90.98 | 5.0 | White & Harlow (1993) |
| September 10, 1978 | 14.27 | 91.50 | 6.0 | NEIC |
| January 12, 1979 | 14.29 | 91.53 | 6.1 | NEIC |
| October 9, 1979 | 14.66 | 90.28 | 5.0 | White & Harlow (1993) |
| August 9, 1980 | 15.89 | 88.52 | 6.7 | NEIC |
| September 2, 1980 | 15.91 | 88.29 | 5.3 | NEIC |

Table 6.1. (*cont.*)

| Date | Latitude (° N) | Longitude (° W) | Magnitude | Reference |
|------|----------------|-----------------|-----------|-----------|
| April 27, 1982 | 14.53 | 87.72 | 5.0 | NEIC |
| September 29, 1982 | 14.55 | 89.13 | 5.1 | NEIC |
| September 29, 1982 | 14.49 | 89.12 | 5.5 | NEIC |
| October 31, 1982 | 14.09 | 90.27 | 5.3 | NEIC |
| December 2, 1983 | 14.07 | 91.92 | 7.1 | NEIC |
| May 29, 1989 | 15.79 | 89.95 | 5.0 | NEIC |
| September 18, 1991 | 14.65 | 90.99 | 6.2 | NEIC |
| December 19, 1995 | 15.30 | 90.15 | 5.4 | NEIC |
| July 6, 1997 | 16.16 | 87.92 | 6.1 | NEIC |
| January 10, 1998 | 14.37 | 91.47 | 6.6 | NEIC |
| March 3, 1998 | 14.38 | 91.47 | 6.1 | NEIC |
| February 16, 1999 | 15.65 | 87.13 | 5.2 | NEIC |
| March 29, 1999 | 14.43 | 88.40 | 5.1 | NEIC |
| May 8, 1999 | 14.21 | 91.94 | 6.1 | NEIC |
| July 11, 1999 | 15.78 | 88.33 | 7.0 | NEIC |
| June 9, 2000 | 14.38 | 90.38 | 5.0 | NEIC |
| January 23, 2001 | 14.02 | 90.91 | 5.9 | NEIC |

[a]National Earthquake Information Center (http://www.neic.cr.usgs.gov/neis/epic/epic.html).

Of particular interest was the damage noted to Monuments 8 and 10 (M8 and M10 in Figure 6.7) in the Great Plaza. These stelae had been found fallen by Maudslay in 1883 and in 1934 they were repaired by Stromsvík with concrete and medially embedded reinforcing rods; this was part of a restoration effort under the aegis of the Carnegie Institution of Washington. Ground shaking from the 1976 event fractured the monument shafts at ground level producing cracking, spalling, and exposure of the reinforcing rods (Figure 6.9). A member of the archaeological team who was on site during a strong aftershock also observed that Monument 8 was rocking in an east–west direction. In addition, bracing of Monument 10 was needed after the earthquake.

What can be inferred from the direction of shaking of Monument 8? Monument 8 has its short axis facing or roughly parallel to the strike of the Motagua fault zone. The Motagua fault is a left-lateral strike-slip fault (the opposite side of the fault moves to the left). In the vicinity of Quiriguá the Motagua fault zone strikes approximately N 62° E. Most of the damage features produced by earthquakes are a result of horizontal shaking from transverse or shear seismic waves. For a left-lateral strike-slip fault the pattern of ground displacements for transverse waves is quadrantal with the maximum lobes of shear displacement parallel and perpendicular to the fault zone. The sense of initial strong ground

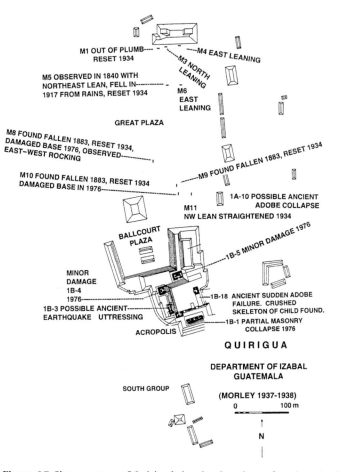

**Figure 6.7** Site core map of Quiriguá showing locations of earthquake damage and leaning stelae. From Morley (1937–1938).

motion depends on the location of a site relative to the earthquake epicenter. Figure 6.10 illustrates the predicted shear wave radiation pattern for the 1976 Guatemala epicenter. Quiriguá, located southeast of the epicenter, lies in the quadrant for which the initial strong ground motion is northeasterly–southwesterly, or parallel to the trend of the Motagua fault.

What can be said about the effects of earlier earthquakes at Quiriguá? The San Agustin, Motagua, and Polochic fault zones have undoubtedly been subjected to earthquake movements during the Holocene. Our ability to recognize an earthquake occurrence thousands of years ago, if it is not documented in some written record, must rely on archaeology. Damage to structures depends on a number of factors such as the magnitude and severity of the earthquake, the geographic location, and the type of construction, building design, and materials used. In

(a)                                        (b)

(c)

**Figure 6.8** Earthquake damage from the 1976 Guatemala earthquake observed at Quiriguá (Bevan and Sharer, 1983). (a) Left-lateral offset of 0.8 m in a stockyard fence on the north side of the ruins. (b) and (c) Earthquake damage to masonry walls in Structure 1B-1 reconstructed during restoration undertaken in 1912.

addition, the nature of the underlying rock or soil plays an important role. Did the eighth-century Maya population at Quiriguá actually experience the tremors associated with crustal earthquakes?

Past earthquake activity was indicated by building damage revealed in excavations of the 1970s. Sudden failure of Structure 1B-18, an ancient adobe building on the east side of the Acropolis, apparently took place near the end of the occupancy of Quiriguá. The building collapse sealed a find of pottery kitchen storage

(a)                                    (b)

(c)

**Figure 6.9** Earthquake damage to Monuments 8 (a) and 10 (b) at the Quiriguá ruins from the 1976 Guatemala earthquake. (c) Masonry buttressing placed against the west wall of a platform in the Acropolis, most likely placed there as reinforcement from continuing earthquake activity (Bevan and Sharer, 1983).

vessels. The structure was neither previously cleared nor rebuilt. Furthermore, it was discovered that many of the early large-vaulted structures were heavily buttressed. Such buttressing, and the evidence of ancient masonry destruction and subsequent repairs, strongly suggest that earthquake activity was a recurring problem for the residents of Quiriguá.

One of the observations that may have a bearing on past earthquake occurrences is the tipping of the stelae in the Great Plaza. Many of these stelae are distinctly leaning and were originally discovered fallen or noticeably out of plumb. In 1840, Stephens observed that Monument 5 had a northeast lean (Stephens, 1841). The monument fell in 1917 (as a result of heavy rains?) and was reset in 1934. Monument 9 was found fallen in 1883 and was uprighted in 1934. Two

## Shear wave radiation pattern at Quiriguá

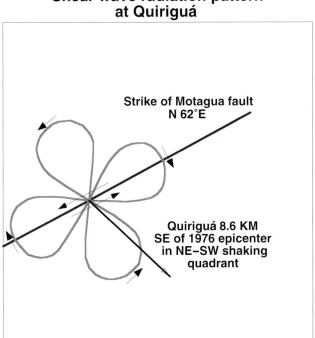

**Figure 6.10** Shear wave radiation pattern for the 1976 Guatemala earthquake located on the Motagua fault. The quadrants of maximum shear displacement are shown.

of the monuments (numbers 8 and 10), which fell in ancient times, were reset and reinforced in 1934 but were subsequently damaged during the 1976 earthquake. Two stelae, Monuments 4 and 6, remain distinctly out of vertical with an eastward lean.

It has been previously inferred that the east–west lean of these monuments is indicative of shaking parallel to the Motagua fault zone, which is located just north of the ruins. If this is true, what is the level of ground acceleration needed to topple a typical Maya stela? Many studies have examined the accelerations needed to topple columns, chimneys, tombstones, Japanese stone lamps (*ishidoro*), and precariously balanced rocks. Toppling accelerations ranging from 0.1 to 0.5 $g$ are not uncommon. The Maya stelae are not free-standing monuments but are placed with a portion of their butts buried beneath the ground as a foundation. Nevertheless, we can estimate the level of ground acceleration, and hence the seismic intensity, needed to topple a free-standing stela. A much higher level of intensity would of course be needed to overthrow a monument with a percentage of its height imbedded in the ground.

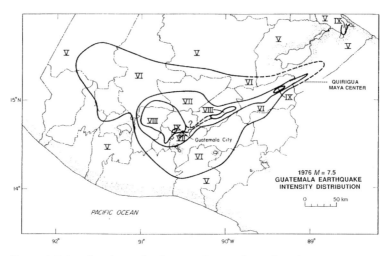

**Figure 6.11** Isoseismal map for the 1976 Guatemala earthquake.

Let us consider a free-standing monument standing on its end and rectangular parallelepiped in shape. The maximum acceleration, $a$, applied at the base, needed to overthrow the body (the overturning moment) is given by the expression:

$$a = xg/y$$

where

$a$    is the maximum horizontal acceleration applied at the base

$g$    is the acceleration due to gravity

$x$    is the horizontal distance from the edge about which the body was overturned

$y$    is the height of the center of gravity above the base

Assume a free-standing solid stela 7.6 m high with a square base 1.2 m in lateral dimension. The horizontal acceleration, at the base, needed to topple the monument would be 0.6/3.8 × 980 cm/s² or 154 cm/s², about 16 percent of the Earth's gravitational acceleration.

Looking at the isoseismal map for the 1976 Guatemala earthquake (Figure 6.11), it can be seen that the ruins at Quiriguá were assigned Mercalli intensities of VI to VII, perhaps a somewhat low value considering the proximity of the causative fault and the presence of surface ground rupture near the site. Nevertheless, it can be deduced that local ground accelerations of 18% to 34% $g$ or Modified Mercalli intensities of about VII would certainly be needed to overthrow a reasonable sized stela, bearing in mind that the stelae are also partially set below ground level. It is likely, however, that such horizontal accelerations could cause rocking motions as were observed during the 1976 earthquake or cause stelae to be aligned out of plumb.

An adobe block building on the east side of the Acropolis (Structure 1B-18) collapsed near the end of the occupation of Quiriguá. A group of pottery kitchen storage vessels and the skeleton of a child were crushed under the fallen walls of the structure. It is not unreasonable to infer that this building failure was caused by an ancient earthquake, which may have dealt a contributing blow to the demise of Quiriguá. Corollary evidence for past earthquake activity is also indicated by the archaeological observation that many of the vaulted structures at Quiriguá were buttressed and the fact that evidence for masonry repair was found. The exact date for the cessation of the primary building activity of Quiriguá, which might coincide with this earthquake, cannot be dated with precision but it appears likely that it was in the time frame of AD 810–850. Sharer (1978) points out that the final constructions at Quiriguá correspond to the Maya Long Count of 10.0.0.0.0, the tenth baktun, or AD 830.

## 6.4     Evidence from Benque Viejo (Xunantunich), Belize

A small Maya site, Benque Viejo, also known as Xunantunich, lies just east of the border with Guatemala (Figure 6.12). Benque Viejo appears to be an eastern member of the Maya sites in the Petén district of Guatemala that include the familiar sites of Tikal and Uaxactun. Benque Viejo was constructed on well-drained limestone with good soil cover that now supports a rain forest. The site is believed to have been a small, provincial ceremonial center. It is dominated by a large building complex, Structure A-6, that occupies one of the highest points in the area (Figure 6.13). One stela, located in front of Structure A-1, has been deciphered to give a date of November 30, AD 849 (10.1.0.0.0 5 Ahau 3 Kayab in the Maya Long Count). Many Maya scholars today do not believe that Benque Viejo lies in earthquake-prone country. However, some rather convincing evidence of structural damage revealed from archaeological excavations in 1959–1960 (Mackie, 1961, 1985) and its proximity to the Caribbean plate boundary would argue otherwise. Let us look at the archaeological evidence.

Benque Viejo consists of a number of masonry structures around the periphery of two main plazas, A and B. Two structures are of importance for our discussion. Structure A-15 was a five-roomed vaulted masonry building believed to have been used by a member of the site's governing class. Dating of the construction of Structure A-15 from shards found in the vicinity suggests a ceramic age of early Benque Viejo IIIb. Benque Viejo IIIb lasted from AD 700 to 830 or 890. The uncertainty in the end date is the result of debate about the exact

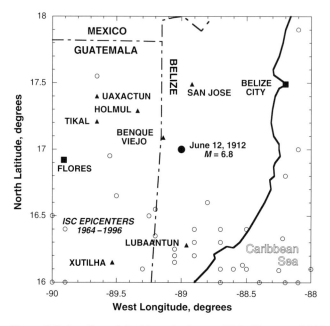

**Figure 6.12** Location of the Maya site Benque Viejo (Xunantunich) just east of the Guatemala border in Belize (Mackie, 1985). Open circles represent recent earthquake epicenters. The solid circle indicates the location of a magnitude 6.8 earthquake on June 12, 1912. ISC, International Seismological Centre.

chronological end of the Classic culture at Benque Viejo based on ceramic parallels and correlations with the nearby Petén site of Uaxactun. Recent studies of the ceramic complexes using radiocarbon techniques (Le Count *et al.*, 2002) favor a date close to AD 890. The last dated stela found at Uaxactan carries the date AD 889. Under the common assumption that stelae were erected to dedicate new buildings, this date would mark the end of any significant building activity at Uaxactun, and by association Benque Viejo. Structure A-15 was severely damaged at the end of the Benque Viejo IIIb period. A sudden fall of the outer walls, which collapsed together with the outer halves of the roof vaults, was indicated from the positioning and the found condition of the masonry debris. There was no evidence to indicate that any attempts were made to reconstruct the building.

Structure A-11 is located on the northern end of the buildings that make up group A. This structure, believed to be the royal residence, consisted of a lower gallery of three rooms and an upper gallery with a minimum of five rooms. When Structure A-11 was first examined it was observed that all of the roof vaults and outer walls of the structure had collapsed, and that the rooms were filled with masonry debris and roof blocks. What was unusual about the discovery

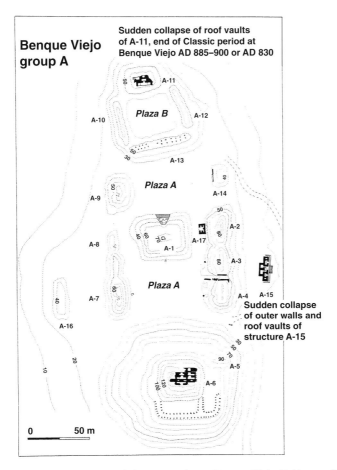

**Figure 6.13** Site map of the Maya ruins at Benque Viejo. Evidence of sudden earthquake collapse was found at Structures A-11 and A-15 (Mackie, 1985).

was the observation that the fallen vault blocks were well preserved, with their plastered faces in near perfect condition. The presence of masonry rubble itself implies that the building fell suddenly. The pattern of broken fragments of vessels found on the floor indicates that some pottery, dated in the style of Benque Viejo IIIa, was standing in the building and was broken in situ by fallen vault stones. All in all the evidence argues that the fall of Structure A-11 was sudden:

(1)    the fresh, found condition of the interior plastered surfaces;
(2)    the lack of dust found on the interior floors;
(3)    the presence of air spaces between the masses of the fallen vault blocks;
(4)    the pattern of broken pottery vessels.

The characteristics and pattern of the observed damage is striking. Recent studies (Le Count *et al.*, 2002) found no evidence of violence and very little evidence of civil disruptions. There were no signs found of any attempts at repair or reconstruction. The pottery found inside, and the lack of floor dust indicate a sudden interruption of occupancy due to collapse while the structure was in a good state of repair. Taken in its totality the evidence argues for abandonment after a sudden earthquake collapse around AD 890.

### 6.5    Were there earthquakes at Copán, Honduras?

The lowland Maya site of Copán was a major Classic-Period ceremonial center located in western Honduras. Copán was an important astronomical center and was highly organized in an architectural sense. Its physical layout comprised a northern Great Plaza and a southern series of rising courts known as the Acropolis Group (Figure 6.14). Copán flourished during the Classic Period from AD 250 to AD 900. Tectonically speaking, Copán lies in the broad diffuse zone of seismic activity marking the western landward extension of the Caribbean plate boundary (Figure 6.15). The rocks underlying the site are volcanic pyroclastic rocks of rhyolitic and andesitic composition that have become consolidated through the percolation of water. Such rocks, known as tuff, are commonly used as a building stone in volcanic areas because they are soft enough to be hand-quarried and of sufficient strength to be set into walls with mortar. The carved stelae found at Copán are of andesitic composition with hard, dense inclusions or nodules which were sometimes left as part of the carved design itself.

It is quite likely, given its geographical position and the location of recent seismic activity, that Copán would have been subjected to earthquakes in the historical past. On December 3, 1934, an earthquake of estimated magnitude 6.2, with an epicenter near Copán, caused great damage in western Honduras. The New York Times of December 8, 1934, reported that 11 towns, including Copán, were destroyed and that "many bodies were found among the ruins of ancient Mayan monuments." Aftershocks were felt for many days. An area of at least 2600 km$^2$ was subjected to a Modified Mercalli intensity of VI and greater, which was sufficient to cause fright, cracking, and damage to weak masonry structures (Figure 6.16). Horizontal ground accelerations at a minimum of about 9 percent of the Earth's gravitational acceleration were reached. What happened at the ruins of Copán? As part of a program of investigation and repair beginning in 1935, workers of the Carnegie Institution of Washington made several relevant and pertinent observations.

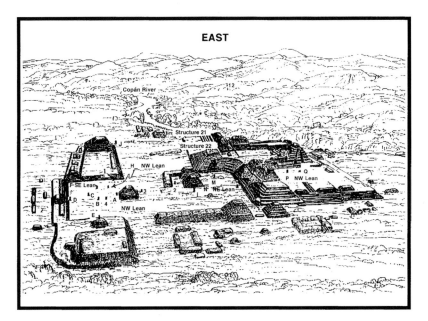

**Figure 6.14** A diagrammatic view of the ruins of Copán, Honduras, looking east. The location of the main stelae and Structures 21 and 22 of the Acropolis are marked. Diagram annotated from Morley (1920).

Stromsvík (1935) stated that:

> An excavation was started under the base of Stela A, on the east side of the plaza. This stela was leaning so badly, apparently as result of a recent severe earthquake, that it was in danger of falling, as were Stelae H and P. [Stelae H and P were described as standing in a dense forest when seen by Stephens in 1839.] One of the first steps was therefore to secure these stelae with stout wooden braces until they could be straightened and their bases strengthened. Earthquakes occurred almost daily, in stronger and weaker degrees.

Structure 22 was in a partial state of preservation prior to the 1934 earthquake. Several of the temple rooms of Structure 22, excavated by the Peabody Museum Honduras Expedition of 1895, caved in and were buried under the fallen walls of the building (Blom, 1935; Trik, 1939). Fissures were formed in the top of the east side of the Acropolis complex and some masonry slid down into the river. Figure 6.17 shows some typical damage from the 1934 earthquake to one of the rooms in Structure 22. The neighboring village of Copán was very badly damaged. One-quarter of its adobe houses, together with its church, collapsed.

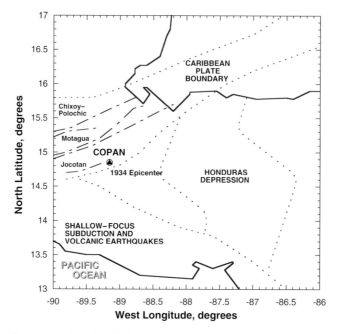

**Figure 6.15** Location of the Maya site of Copán and the epicentral location of the $M = 6.2$ earthquake of 1934. Broadly speaking Copán lies in the diffuse landward extension of the Caribbean plate boundary.

Saville (1892) visited Copán in December of 1891 as part of the Peabody Museum Honduras Expedition; he unequivocally described his observations:

> There is another long crack in the Main Structure running from north to south parallel with the river front. The walls in nearly all the buildings explored in Copán have fallen outward presenting a different appearance from those of the ruined edifices in Yucatán. While the buildings in Yucatán were probably destroyed by vegetation aided perhaps by the hand of man, in Copán we have evidence of a stronger force, probably an earthquake, producing the long cracks seen in the Main Structure, breaking off several idols just above the base, and wrenching others from their foundations . . . The roots of trees have done much to destroy the upper parts of these massive steps [the steps of mound 21 at Copán], but the lower part are in an excellent state of preservation, and it seems very probable that the building which once stood above, fell before any vegetation grew upon the structure.

Saville, in his investigations, further reported that Stelae A, B, D, F, H, J, N, and P were standing in 1892 and that Stelae C, E, I, M, 1, 2, 3, 4, and 12 (Figure 6.14) were fallen and broken. It is useful to consider some of these observations in

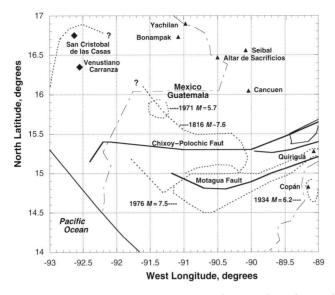

**Figure 6.16** Areas of intensity VII or greater for some damaging earthquakes of Guatemala. Copán was subjected to a Modified Mercalli intensity of at least VII during the 1934 event.

greater detail. In 1839, Stephens found Stela C broken into two large pieces, and Stela M fallen and ruined. One would assume that these stelae were broken at some unknown time prior to 1839. Stela I is located on the east side of the Middle Plaza and was described by Stephens as standing but covered over with masonry and earth. Stela I was subsequently found by Maudslay in 1885 with its upper part broken. Stela 4, one of the most striking of all of the structures at Copán, is broken off just above the base; it was described by Stephens as fallen and lying on its side. Stela 10, located on the summit of a hill about 4 km west of the ruins, lies flat on the ground. Stela 12, located east of the Copán River, is fallen and broken. Both of these stelae were apparently aligned or positioned to mark some astronomical event in the past and were first reported by Galindo in 1835. Nine other stelae, located west of the Main Structure and not shown in Figure 6.14, were also reported fallen or shattered.

Taken in their totality the number of fallen stelae is supportive of Saville's (1892) contention of earthquake occurrences at Copán prior to his visit and those of Maudslay in 1881–1885. One, of course, cannot totally exclude the possibility that native clearing and felling of timber in the 1860s could have toppled and damaged some of the monuments after Stephens' visit in 1839; Stephens repeatedly emphasized that Copán at that time was enveloped in a dense tropical forest. Archaeological excavations carried out in 1983 uncovered a collapsed masonry building in one of the residential compounds of the nobility. Inside

(a)

(b)

Figure 6.17 Damage (a) and reconstruction (b) to the west end of the south room in Structure 22 at Copán (courtesy of the Carnegie Institution of Washington).

the building, artifacts and partly finished objects were found in their original positions on a bench; the floor was covered by a heavy beam-and-mortar roof that had suddenly collapsed, most probably as the result of an earthquake. Archaeologists dated this event as taking place about AD 950–1000 (Webster, 2002). An additional factor arguing for past earthquake damage is the proximity of Copán to Quiriguá, where earthquake damage to stelae has been witnessed

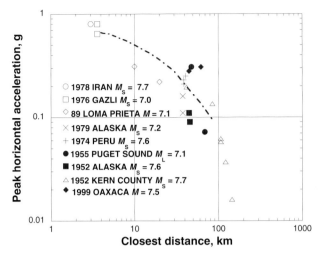

**Figure 6.18** Plot of peak horizontal ground acceleration versus closest distance to causative fault for some earthquakes in the magnitude range 7 to 7.6.

and documented at levels of ground shaking appropriate for Modified Mercalli Intensities of VII and VIII. Such levels of ground accelerations, 18% to 34% g, are not inappropriate in the meizoseismal area of earthquakes of magnitudes greater than 7 (Figure 6.18). Therefore, it seems quite likely that earthquakes could easily have affected Copán in the historical past.

## 6.6    Seibal, Guatemala

Seibal is a southern lowland Maya site in the Rio Pasión region of the Petén district. Latitude and longitude coordinates are approximately $16°\,30'$ north and $90°\,05'$ west. The stelae at Seibal indicate a vigorous occupation in the time interval from AD 830 to 890 at a time when other Maya centers in the Petén were gradually declining in importance. Shortly after AD 900 Seibal was abandoned.

Seibal is positioned in a marginal fold belt close to the landward extension of the Caribbean plate boundary zone (Figure 6.2). Areal geology is predominantly horizontally bedded marls and intercalated limestone strata of Permian and Cretaceous age. The surface topography is dominated by parallel ridges that strike predominantly north–south with downthrown, sometimes water-filled, fault blocks on the sides. These graben-like structures are most probably tensional features induced by left-lateral strike-slip movement on the Polochic and Motagua fault systems. These fault systems pass east–west through central Guatemala turning to a bearing $N\,60°\,E$ just south of Seibal.

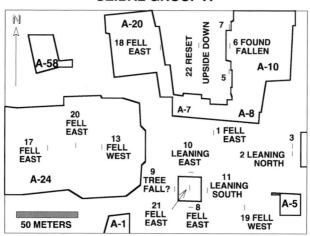

Figure 6.19 Generalized site map of structures in Group A at Seibal, Guatemala, showing location of fallen, but now repaired and erect, stelae.

The locations of earthquake epicenters in this area of Mexico, Guatemala, and Belize demonstrate that tectonic activity is currently taking place and there is no reason to believe that Seibal in its prime did not also experience earthquakes. The possibility of any earthquake effects at the site, however, seems to have been discounted by archaeological investigators (Graham, 1990). When Maler (1908a) visited and photographed the site in 1895 he noted that many of the stelae were fallen and broken (Figure 6.19). Later archaeological expeditions to Seibal were conducted by the Peabody Museum, Harvard University from 1964 to 1968; 15 of the stelae were mended and reset (Willey et al., 1975). Nine stelae at Seibal (numbers 1, 8, 13, 14, 17, 18 to 21), first seen in the 1890s, were found fallen in either an east or west direction. The explanation offered by early investigators was that the stelae were toppled by trees, either naturally or by later woodcutters. However, does it not seem more than random that 83 percent of the stelae found fallen were lying in an east or west direction?

Seibal lies in a seismically active area, the diffuse landward extension of the Caribbean plate boundary, therefore east–west shaking at the site would not be surprising. On an east–west striking fault plane, movement in a *left-lateral* sense would produce the maximum values of shear displacement at bearings 90° and 270° from north. As a result the maximum shaking or horizontal accelerations would be expected to topple or rock monuments preferentially in these directions. Such east–west rocking of stelae was witnessed by observers during earthquakes in 1976 at Quiriguá. If current seismicity is a measure of seismic activity of the past, then earthquakes cannot be excluded as a plausible contributor to damage to the monuments at Seibal. It is interesting to speculate whether the suggested earthquake damage was produced by a single earthquake or several, and when an earthquake might have occurred. Firm answers cannot be given although any occurrence would undoubtedly have taken place after the last stela at Seibal was erected. The upper fragment of fallen Stela 6 – Stela 22 – was discovered in close proximity, but *reset upside down* by unknown parties, sometime after AD 889; such a condition suggests that the initial damage to the stelae occurred near the time of the abandonment of Seibal.

## 6.7    Altar de Sacrificios and Xutilha, Guatemala

Altar de Sacrificios is located 60 km west of Seibal on the south bank of the Rio Pasión, in the southwestern part of the Department of Petén, Guatemala. The site was not as large as Seibal and its architectural remains are not extensive but it is important because its stelae were carved from both sandstone and limestone. After 9.10.0.0.0 (AD 633) no sandstone stelae can be found, indicating that there was a change in the type of architectural stone used. The site was apparently invaded and conquered near the close of the Classic Period (in about AD 900).

As in Seibal, Altar de Sacrificios must also have been subjected to earthquakes with epicenters in the Caribbean plate boundary zone. Early investigators, such as Maudslay, Maler, and Morley gave various partial descriptions of the site but in-depth archaeological excavations did not begin until the late 1950s (Graham, 1972; Willey and Smith, 1969; Willey, 1973). Many of the stelae in one of the monumental groups at Altar de Sacrificios were found fallen (Table 6.2).

Xutilha is a recently discovered ruin in the southeast portion of the Petén district, first investigated in 1957 (Satterthwaite, 1961). Its approximate coordinates are 16.19° N and 89.53° W which place it about 80 km southeast of Seibal and only 40 km north of the seismically active Polochic fault zone. Xutilha is a small site but 12 stelae were discovered, probably carved during Late Classic occupation. Interestingly, as might be expected in an earthquake-prone region, there was no visible manifestation of masonry vaulted roofs for buildings. Of the 12 stelae at Xutilha (Table 6.3), 5 were found fallen in a direction appropriate for transverse shear motion originating on the Polochic–Motagua fault systems located directly to the south.

Earthquake damage, which occurred near the end of the Classic Period, was noted at the Maya sites of Benque Viejo and Quiriguá. Tilted and fallen stelae were found at Copán, Seibal, and Altar de Sacrificios. Could one earthquake have caused this damage? If it were one earthquake how large would it have been? Let us first examine the current pattern of earthquake occurrences in

Table 6.2. *Stela at Group A, Altar de Sacrificios*

| | |
|---|---|
| Stela 1 | Found at base of Structure A-II. Butt found 5 m behind in ruined building. Fell to east. |
| Stela 2 | Found on ground, original position unknown. Butt missing. |
| Stela 3 | Originally faced east. Fallen on back, to west. |
| Stela 4 | Faces south, upright. Thin, probably a wall panel. |
| Stela 5 | Faces south, found upright. One piece, probably a wall panel. |
| Stela 7 | Butt in place. Stela faced east. Fragments lie in front face down, east. |
| Stela 8 | Complete in one piece. Found fallen on back imbedded in ground to half its thickness indicating fallen for some time after abandonment. Direction not given. |
| Stela 9 | Fallen but complete in one piece. Direction not given. Original positioning uncertain. |
| Stela 15 | Butt in place, faced north. Fell south, 13 fragments. From a heavy blow? |
| Stela 16 | Standing, leaning heavily south and east. Substantial and well-planted butt. Strong force required to produce lean. |
| Stela 17 | Faced north, butt in position. Eight fragments reassembled in 1944. No mention of orientation of fragments when found. |
| Stela 18 | Found fallen, front up (north?), completely buried. Butt missing. Probably moved. |

Table 6.3. *Stela at Xutilha*

| | |
|---|---|
| Stela 1 | Butt *in situ*, few fragments found. |
| Stela 2 | Butt fragment. |
| Stela 3 | Erect, height 1.2 m. Possibly re-erected. |
| Stela 4 | Fell backward (easterly). Height 2.7 m. Badly shattered. |
| Stela 5 | Butt *in situ*. |
| Stela 6 | Fell backward (east-southeast). Cracked in two near base. Height 4 m. |
| Stela 7 | Butt *in situ*. Few fragments. |
| Stela 8 | Fell backward in one piece (south-southeast). Height 3.5 m. Width 1.3 m. Thickness 0.45 m. Acceleration of 15% g could topple. |
| Stela 9 | Standing unbroken. Height 1.2 m. Leans forward (westerly). |
| Stela 10 | Upper portion broken into two fragments. Butt fragment leans forward and to left (northwesterly). Height 1.0 m. Possibly reset. |
| Stela 11 | Fell backward face up and to left (easterly). Height 2.0 m. Width 1.2 m. Thickness 0.25 m. Acceleration of 15% g could topple. |
| Stela 12 | Slumped backward (southeasterly) cracked into two fragments. Possibly re-erected. |

the region in question. Figure 6.20 shows earthquake epicenters located by the International Seismological Centre (ISC) and the Natural Earthquake Information Center (NEIC) for an area between longitude 86° and 92° W. The time interval is from 1964 to 1999. Magnitudes are not discriminated but range from about 3.7 to 7.5. Much of the seismic activity is associated with the Motagua–Polochic

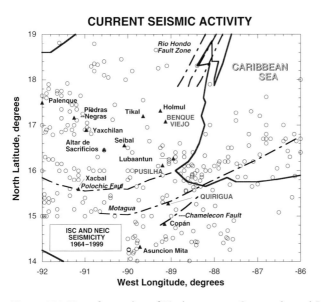

**Figure 6.20** Map of a portion of Mexico, eastern Guatemala, and Belize showing earthquake epicenters from 1964 to 1999. Locations of the Polochic, Motagua, and Chamelecon faults are shown, together with the locations of various Maya sites. Epicenters are shown by open circles.

fault system, but diffuse activity is present in the southern lowlands. The number of events decreases as the northern and central lowlands are approached. Assuming that this areal distribution of seismicity is a valid measure of historical activity, it is clear that a large portion of the Maya region has been subjected to felt earthquakes.

Figure 6.21 shows an empirical correlation of the area enclosed by a Modified Mercalli intensity of VII in square kilometers versus magnitude for worldwide earthquakes. Ordinate values are moment-magnitude values when they exist, otherwise the values are local or surface wave magnitude determinations. An intensity of VII was selected as the critical level of shaking because it represents the level at which structures constructed of adobe or masonry, which are weak to horizontal accelerations, are damaged. It also represents the level of ground acceleration for which stelae would either rock or topple. Using data from this plot we might expect an $M = 7.5$ shock to produce an area of about $30\,000$ km$^2$ subjected to an intensity of VII or greater. Likewise, an $M = 7.9$ event would encompass an area of about $80\,000$ km$^2$. To examine the area affected by earthquakes of this size, let us assume a hypothetical epicenter on the seaward extension of the Polochic fault (Figure 6.22). By drawing a circular arc we can show the expected extent of the area shaken at a level of VII by our assumed

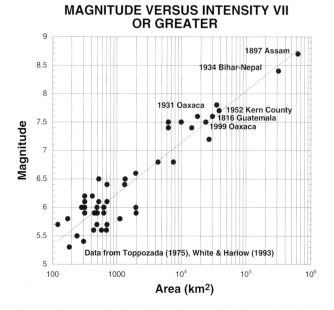

**Figure 6.21** A compilation of data from worldwide earthquakes plotting earthquake magnitude against the logarithm of the area subjected to a shaking intensity of VII or greater on the Modified Mercalli Scale.

earthquake. It can be seen that the Maya sites of Benque Viejo, Quiriguá, Copán, and Pusilhá could easily have been damaged by one or more earthquakes at or near the end of the Classic Period. If the hypothetical epicenter were shifted westward, the site of Seibal could also be included in this zone of potential damage.

However, there is no a-priori reason to argue that a single earthquake caused physical damage and subsequent abandonment of many lowland sites. It is more plausible that earthquakes played a triggering role. Aside from the destruction of buildings and temples, earthquakes may well have exacerbated the disintegration of the social order at several Maya centers in the southeastern lowlands. The priesthood, which formed the hierarchy of the Classic Period, undoubtedly controlled the procurement of food, labor, and services of the agriculturally based population within their realm of influence. Part of this dominance must have been achieved by the belief of the populace that their priests possessed a special ability to communicate with the gods. The sudden destruction of temples and sanctuaries by other-than human causes could have weakened the priestly hold on the population, leading to their refusal to rebuild structures at the site and ultimately leading to abandonment.

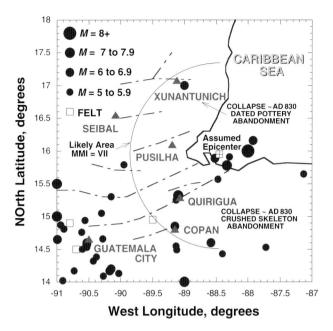

**Figure 6.22** The circular arc represents the western boundary of Modified Mercalli intensity (MMI) VII or greater for a possible $M = 7.9$ earthquake centered as shown at the Caribbean coastline. Solid circles represent location of historical earthquakes of $M = 5$ or greater. Data are given in Table 6.1.

## 6.8    Pusilhá, Belize

Pusilhá is located at 16.1° N, 89.19° W in the southwestern corner of Belize, very close to the border with Guatemala. It is situated only a short distance north of the Polochic fault, a seismically active strand of the North American–Caribbean plate boundary. Pusilhá is an unusual site in that it contained a concentrated group of sculptured stelae but its presence was unknown until it was accidentally discovered in the late 1920s (Joyce *et al.*, 1927).

The report describing the excavations at Pusilhá (Joyce *et al.*, 1928) states that "all [of the stelae] had probably been knocked down and broken by falling trees, which is not surprising when one considers the shallowness of the foundations supporting the monuments, as compared with those of other Maya cities." Most of the stelae had fallen with their carved inscriptions downward so that they were fairly well preserved.

Figure 6.23 is a site map of the main ruins at Pusilhá showing the positions of Stelae A, B, C, D, E, F, G, H, K, L, M, and N, which faced the plaza. The stelae were aligned with a bearing N 23° E. Only the bases for Stelae A, G, and L were found and, except for Stela C, all of the remaining stelae had fallen, inscription-side

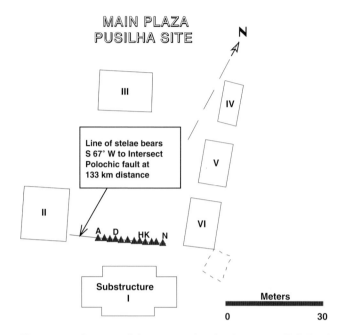

MAIN PLAZA
PUSILHA SITE

Line of stelae bears
S 67° W to Intersect
Polochic fault at
133 km distance

**Figure 6.23** Site map of the excavated Main Plaza at Pusilhá showing the line of
stelae, which had preferentially toppled in a northwesterly–southeasterly direction.

down, in a direction S 23° E (Stelae B, D, E, F, H, K, and N). Stelae C, D, E, H, and
K were subsequently trimmed and cut into smaller pieces and removed to the
British Museum in London.

Compared to the stelae at Quiriguá, the carved monuments at Pusilhá were
smaller, having approximate dimensions of 300 cm in height, 90 cm in width,
and 30 cm in thickness. The stelae were aligned with their widths in a direction
S 67° W. Owing to their thickness, a horizontal acceleration of only 9 percent
of the Earth's gravitational acceleration would have been needed to topple a
free-standing or shallow-rooted stela at Pusilhá. This level of ground acceleration
would easily be achieved with a Modified Mercalli intensity value of VI produced
by a nearby earthquake.

Let us assume that transverse (shear) waves were responsible for producing
the consistent directions of the fallen stelae. The line of stelae, which is at right
angles to the common directions of fall, would represent the azimuth to a hypothetical epicenter on a causative fault. As shown in Figure 6.23, this azimuth
of S 67° W intersects the Polochic fault, a major strand of the North American–
Caribbean plate boundary, at a distance of 133 km. A magnitude 7.5 earthquake
at this distance would easily produce a Modified Mercalli intensity value of VI
and ground accelerations of 9% to 10% g. The expected recurrence rate for magnitude 7.5 earthquakes along this plate boundary is about 100 years. Other events

may have affected Pusilhá and the possibility of earthquake damage cannot be dismissed. Without stretching the long arm of coincidence, the preferential directions of fall argue in favor of earthquake-induced damage.

The period of occupancy of Pusilhá is uncertain by several katuns in the Maya Long Count. The earliest conclusive inscription is dated 9.7.0.0.0 7 Ahau, 3 Kankin (December 5, 573). It is necessary to assume a somewhat earlier date for the actual foundation of the site – at least several katuns, or approximately AD 533. The last conclusive inscribed date, found on Stela E, was 9.15.0.0.0 4 Ahau, 13 Yax (August 20, 731). By sequencing the monuments, and making use of the esthetic quality of the carvings on the stelae for which the dating glyphs had been obliterated, it was inferred that monument construction ceased around 9.18.0.0.0 or AD 790 (Morley, 1938). It is unlikely that the entire population would have deserted Pusilhá immediately after the erection of the last dated stela but it is not implausible that abandonment may indeed have been influenced by a great earthquake.

7

# Earthquakes of Costa Rica, Panama, and Colombia

## 7.1    Legends and early history

The Bribri are an aboriginal people whose homelands are in the eastern river valleys that drain into the low-lying Caribbean coastal areas of southern Costa Rica and northern Panama (Figure 7.1). They were hunter-gatherers and their habitat is somewhat removed from the main axis of Central American volcanic activity. The Bribri were also distant from the main zones of frequent earthquake activity along the Pacific coast but are positioned in a region that borders the Caribbean side of Costa Rica and Panama and that is occasionally subjected to major earthquakes, such as the events in 1822, 1904, and 1991. As a result, legends and stories pertaining to earthquake occurrences are not numerous. In the oral traditions of the Bribri, *Sibú* is the creator god who made the world of rock but also wanted to make the earth so that he could create mankind. A legend of note refers to the occurrence of an earthquake that the god Sibú produced to shake the forested regions:

> vieron una gran luz, una cara roja y brillante como de fuego que las miraba desde la copa de un árbol. Lanzaron un grito; lanzaron diez gritos . . . fue tan terrible el grito, tan estridente que Sibú se estremeció, el árbol tembló, la tierra y el bosque también temblaron . . .

> A great light was seen, a red face, brilliant like the fire that is seen from the top of a tree. There was a scream, ten screams . . . followed by a very terrible scream, so loud that Sibú himself shook, the tree trembled, and the ground and the forest was also shaking. (Peraldo and Montero, 1994)

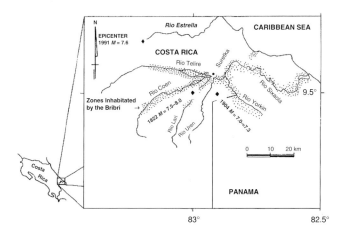

**Figure 7.1** Areas occupied by the Bribi in the Caribbean coastal areas of Costa Rica and Panama. Epicenters of several major earthquakes are shown.

In examining the seismic history of Costa Rica and Panama it is to be noted that these regions were not colonized until 1538–1563, later than Mexico, Peru, Guatemala, and Colombia. In 1508, Spanish settlements were made on the coast of Colombia. In 1536, expeditions proceeded inward from the coast, culminating in the founding of Bogotá in 1538. The early inhabitants of western Colombia did not build with stone and therefore there are no ancient archaeological ruins. Historically the written seismic record for Colombia extends back to 1575.

Cartago, in Costa Rica, was founded in 1563 but no buildings of permanent significance were erected in the sixteenth century. Colonial documents report that earthquakes occurred a number of times from 1638 to 1821 causing damage in Cartago, principally to adobe churches (Table 7.1). Such damage is suggestive of Modified Mercalli intensity levels of VI to VII. The earthquake(s) of 1781 also produced damage to the first adobe church constructed in the city of San José in 1776.

## 7.2    Tectonic and seismic setting of Costa Rica and Panama

From time to time, Costa Rica and Panama are subjected to large, destructive earthquakes. Their geographical position in relation to the tectonic plates of the region is shown in Figure 7.2. Costa Rica and Panama are positioned north of the triple intersection between the Cocos, Nazca, and Caribbean plates. Most of the destructive inland Central American shocks occur within the Caribbean plate. The senses of plate motion are primarily subduction along the Pacific coast Middle America Trench; this motion changes to horizontal strike-slip or transform motion along the northern boundary of the Nazca plate. In

Table 7.1. *Historical damaging earthquakes in Cartago, Costa Rica*

| Date | Epicenter | Inferred Modified Mercalli intensity |
|------|-----------|--------------------------------------|
| 1638 | | VII |
| 1678 | | VII |
| 1680 | | VII–VIII |
| January 1715 | | VI+ |
| September 17, 1727 | | VI+ |
| March 26, 1728 | | VI+ |
| 1746 | | VI+ |
| July 14, 1756 | | VII–VIII |
| February 15, 1772 | 10.1° N, 84.1° W | VII–VIII |
| June (?), 1781 | | VII |
| May 4, 1794 | | VI |
| February 21, 1798 | 10.2° N, 82.9° W | VI |
| December 27, 1803 | 8.1° N, 83° W | VI+ |
| 1811 | | VI |
| 1821 | 10° N, 85.1° W | VII |

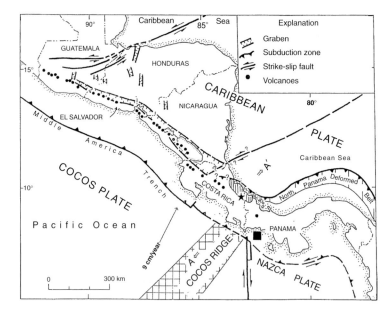

**Figure 7.2** Tectonic setting of Costa Rica and Panama. Location of cross-section A–A′ is shown. The location of the $M = 7.6$ earthquake of April 19, 1991 is shown by a star. The location of the $M = 7.6$ Panama earthquake of July 18, 1934 is shown by a box. The hatched area represents the aftershock area of the 1991 earthquake (Plafker and Ward, 1992).

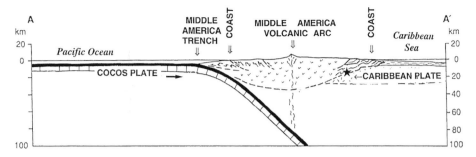

**Figure 7.3** Cross-section through A–A′ in Figure 7.2 (from Plafker and Ward, 1992). The star marks the hypocenter of the 1991 event.

Costa Rica, convergence of the Caribbean and Cocos plates occurs in a direction towards the Middle America zone of volcanic activity. The area offshore to the north of Panama and a part of Costa Rica, called the north Panama deformed belt, exists as a zone of oblique convergence between southern Central America and the Caribbean plate. A diagrammatic cross-section A–A′ across the zone of convergence of the Cocos and Caribbean plates is shown in Figure 7.3.

An unknown aspect of the plate convergence in this area is the partitioning of movement, between the Pacific side and the Caribbean side, that is being accommodated on the opposite sides of the Middle America Volcanic Arc. The Pacific side, known as the forearc side, undergoes motion by subduction of the Cocos plate but an unknown fraction of movement must also be accommodated by occasional significant earthquake motions on the Caribbean (backarc) side. Over the past 100 years, approximately $7.7 \times 10^{25}$ dyne cm of seismic moment has been released per year in the Pacific Coast subduction zone earthquakes of Costa Rica (Table 7.2 and Figure 7.4). Assuming a subduction zone length of 200 km, a downdip width of 35 km, and $\mu = 3 \times 10^{11}$ dyne/cm$^2$, one obtains a *seismic slip rate* of 3.7 cm per year. Since the Pacific plate convergence rate in this region is approximately 9 cm per year this leaves a slip deficit of ~5 cm per year which has to be accommodated by motions on the backarc (or Caribbean) side and/or in the Middle America intraplate region. This calculation emphasizes that significant earthquakes ($M \simeq 7.6$) can be expected to occur occasionally on the Caribbean side of Costa Rica and along the north Panama deformed zone.

To summarize, damaging shallow-focus to intermediate-focus earthquakes are associated with the northeasterly dipping subducting Cocos plate. In addition, shallow-focus earthquakes are also aligned along the Central American volcanic axis or front. These particular earthquakes do not correlate well with individual volcanic eruptions; this suggests that regional plate tectonic stresses, rather than the movement of magma, controls the occurrence of these shocks at depth. Finally, earthquakes are associated with backarc movement along faults

Table 7.2. *Some significant earthquakes of Costa Rica*

| Date | Latitude ($^{\circ}$ N) | Longitude ($^{\circ}$ W) | Depth (km) | Magnitude |
|------|------|------|------|------|
| April 3, 1910 | 9.85 | 83.92 | | 5.8 |
| May 4, 1910 | 9.85 | 84.02 | | 6.4 |
| April 24[a], 1916 | 11.00 | 85.00 | 60 | 7.6 |
| October 5[a], 1950 | 10.35 | 85.20 | 60 | 7.7 |
| December 30, 1952 | 10.30 | 83.50 | | 5.9 |
| September 1, 1955 | 10.25 | 84.25 | | 5.8 |
| April 14, 1973 | 10.60 | 84.70 | 33 | 6.7 |
| February 28[a], 1974 | 9.34 | 84.06 | 46 | 6.2 |
| August 23[a], 1978 | 10.21 | 85.25 | 56 | 7.2 |
| July 3, 1983 | 9.65 | 83.69 | 33 | 6.7 |
| October 4[a], 1987 | 10.82 | 85.93 | 47 | 6.2 |
| March 25[a], 1990 | 9.81 | 84.83 | 26 | 6.4 |
| March 25[a], 1990 | 9.92 | 84.81 | 22 | 7.0 |
| December 22, 1990 | 9.87 | 84.30 | 17 | 6.1 |
| March 1, 1991 | 10.94 | 84.64 | 196 | 6.1 |
| March 16, 1991 | 10.17 | 85.18 | 33 | 6.3 |
| April 22, 1991 | 9.69 | 83.07 | 10 | 7.6 |
| April 24, 1991 | 9.74 | 83.52 | 12 | 6.2 |
| March 4, 1992 | 10.21 | 84.32 | 78 | 6.2 |
| July 10, 1993 | 9.80 | 83.60 | 20 | 5.6 |
| September 4[a], 1996 | 9.36 | 84.27 | 32 | 6.2 |
| August 20[a], 1999 | 9.04 | 84.16 | 20 | 7.0 |

[a]Pacific coast subduction earthquake. Earthquake magnitudes converted to seismic moment using relation $M = 2/3 \log M_0 - 6$, where $M_0$ is moment in newton meters ($10^{-7}$ dyne cm).

associated with the north Panama deformed belt. These faults dip landward from the Caribbean Sea beneath Costa Rica and Panama.

Significant surface deformation and a small tsunami accompanied the destructive $M = 7.6$ Costa Rica–Panama earthquake of April 22, 1991. An area of $15\,000$ km$^2$ was affected with nearly $10\,000$ people left homeless. Much of the widespread damage was produced because the earthquake struck in the coastal lowland region that was underlain by thick unconsolidated sedimentary deposits. Liquefaction and lateral spreading of the surficial materials was common. In addition to the 1991 event, other events have taken place in historical time along this section of the Caribbean coast (Figure 7.5). Details of many of these events are not known; however it is known that rupturing occurred in segments of the north Panama deformed belt, which extends east-northeasterly from the epicentral region of the April 1991 earthquake. A significant gap in historical

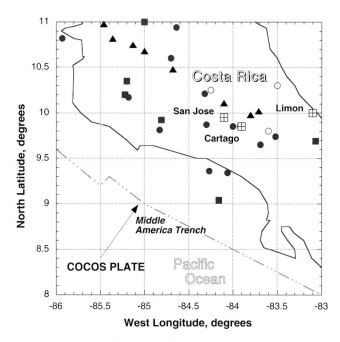

**Figure 7.4** Location of some significant twentieth century earthquakes of Costa Rica. Triangles represent locations of volcanoes. Open circles, earthquakes of $M < 6$; filled circles, $M < 7$; filled squares, $M \geq 7$.

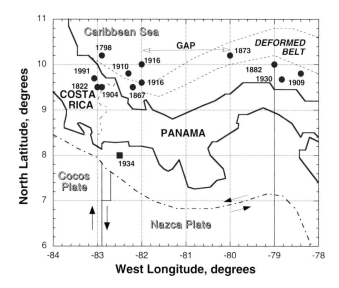

**Figure 7.5** Significant earthquakes of the north Panama deformed belt. Data from Camacho and Viquez (1993).

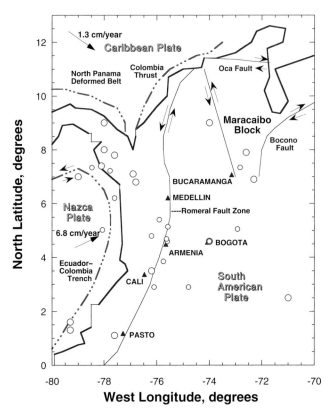

**Figure 7.6** Tectonic setting of Colombia showing plate interactions and major mapped faults. Twentieth century epicenters are shown by open circles. Small circles are $M = 6.5$ to $6.9$; larger circles are $M \geq 7$.

seismic activity is visible in the central segment of the north Panama seismic zone, between the longitudes of $80°$ and $82°$ W.

On the Pacific coast side of Panama, along the northern and western boundaries of the Nazca plate, the sense of motion is primarily left-lateral or strike-slip. Active subduction is not taking place and so earthquakes with magnitudes greater than 7 do not occur very often. A notable exception was the $M = 7.6$ Puerto Armuelles, southwest Panama, earthquake of July 18, 1934.

### 7.3    Seismic framework of Colombia

Colombia is situated within a complex zone of deformation involving three-way interaction between the Nazca, Caribbean, and South American tectonic plates (Figure 7.6). The pattern of surface faulting is very diffuse, emphasizing that the interactions between the plates are spread over a wide zone. Major earthquake occurrences in Colombia are produced both by the effects of subduction and by shallow strike-slip faulting. The subducting Nazca plate moves

Table 7.3. *Twentieth century earthquakes of M ≥ 7 in Colombia and vicinity*

| Date | Latitude (°N) | Longitude (°W) | Depth (km) | Magnitude |
|---|---|---|---|---|
| September 18, 1900 | 4.6 | 74.0 | | 7.9 |
| August 8, 1903 | 4.6 | 74.0 | | 7.7 |
| January 20, 1904 | 7.0 | 79.0 | | 7.8 |
| January 31, 1906 | 1.0 | 81.5 | | 8.8 |
| February 3, 1906 | 3.5 | 76.2 | | 7.6 |
| July 14, 1906 | 4.6 | 74.0 | | 7.6 |
| April 10, 1911 | 9.0 | 74.0 | 100 | 7.2 |
| May 28, 1914 | 9.0 | 78.0 | 70 | 7.2 |
| August 31, 1917 | 4.6 | 74.0 | 100 | 7.3 |
| January 17, 1922 | 2.5 | 71.0 | 445 | 7.6 |
| March 29, 1925 | 8.0 | 78.0 | 60 | 7.1 |
| January 9, 1936 | 1.1 | 77.6 | | 7.0 |
| May 14, 1942 | 0.0 | 80.12 | | 7.6 |
| July 9, 1950 | 7.9 | 72.6 | 41 | 7.0 |
| April 21, 1957 | 6.9 | 72.3 | | 7.1 |
| January 19, 1958 | 1.3 | 79.3 | 40 | 7.7 |
| July 13, 1974 | 7.8 | 77.6 | 12 | 7.3 |
| July 11, 1976 | 7.4 | 78.1 | 3 | 7.0 |
| December 12, 1979 | 1.6 | 79.3 | 24 | 8.2 |
| October 17, 1992 | 6.8 | 76.8 | 14 | 7.0 |
| October 18, 1992 | 7.0 | 76.9 | 10 | 7.4 |

eastward beneath the South American plate and seismic activity associated with this plate goes as far east as Bucaramanga to end in a cluster of repeating intermediate focal-depth earthquakes. Above this subducting plate, the pattern of shallow earthquake activity is scattered and diffuse but many hypocenters appear to be associated with mapped geological features such as the Romeral fault system. In northeastern Colombia a triangular area, the Maracaibo block, moves north and northwestward along the Bocono right-lateral fault system and the north-trending left-lateral Santa Marta–Bucaramanga fault system. Table 7.3 lists the earthquakes of magnitude greater than or equal to 7 that have taken place in Colombia in the twentieth century.

Along the Pacific coast of Colombia the Nazca plate converges beneath the South American plate at a rate of 6 to 7 cm per year. Long segments of the subduction zone sometimes rupture in a single great earthquake, such as the great $M_w = 8.8$ Ecuador–Colombia earthquake of January 31, 1906. The area shaken at a Rossi–Forel level of VII was at least 150 000 km$^2$ (Figure 7.7). A large, local tsunami was also generated, which damaged most of the coastal regions of southwest Colombia and northern Ecuador. This segment of the Nazca plate

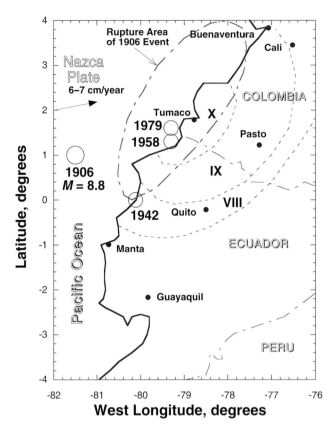

**Figure 7.7** Rossi–Forel intensity distribution for the 1906 Colombia–Ecuador event. The estimated rupture area for the 1906 event encloses rupture zones for 1942, 1958, and 1979 events.

boundary was quiet for 36 years after the 1906 event. Rupture took place within this same segment in 1942 ($M_s$ = 7.9), 1958 ($M_s$ = 7.9), and 1979 ($M_s$ = 7.7), with rupture segments that abutted each other. What is curious, however, is that even though these later events ruptured approximately the same segment of the plate boundary, the cumulative seismic moment release of these events was only 20 percent of the 1906 event. The historical record for Colombia is probably incomplete for events occurring offshore, but does suggest that events comparable in size to the 1906 event are rather infrequent. In any case, failure of smaller segments along a plate boundary segment makes the estimation of the recurrence time of larger events along this segment difficult.

8

---

# Earthquakes of Peru and Chile

## 8.1    Great Peruvian earthquakes

A portion of the western coast of South America within the latitudes of 6° and 19° S is repeatedly struck by great, subduction-zone earthquakes having their epicenters along the Peru Trench. The trench forms the eastern boundary of the Nazca plate and runs parallel to the western coast of South America. These earthquakes typically rupture a large (100 km long, or more) segment of the plate boundary. In addition, these earthquakes often produce strong ground shaking, extensive damage, large tsunami, surface ruptures, landslides, and significant zones of aftershock activity. The coastal regions of Peru and Chile have been subjected in historical times to a significant number of damaging earthquakes. Figure 8.1 shows the areas in Peru subjected to seismic intensities on the Modified Mercalli Scale of VIII or greater as a result of past earthquakes. It can be seen that most of Peru's coastal zone has experienced severe shaking. Intensity data are lacking for the more inland areas because in historical times there were no large European settlements in the interior. Most of the early written records describing earthquake effects in Peru were more concerned with *la gente que cuenta,* the so-called "important people" of European stock. As a result, many of the areas delineated as being subjected to an intensity of VIII may be significantly larger than shown.

Table 8.1 lists some of the great earthquakes that have struck the South American coastal zone since 1582. Assessment of the size of these early events was made from macroseismic observations such as the maximum length of the contour enclosing a Mercalli intensity of VIII (where adobe constructions are severely to completely destroyed) and local tsunami heights. Empirical correlations demonstrate that the maximum height of a tsunami wave (in meters), $H_r$,

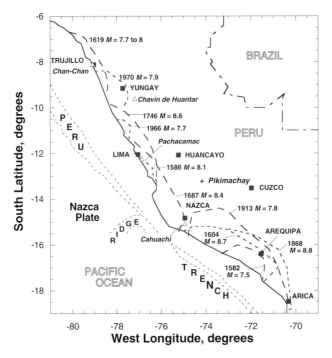

**Figure 8.1** Map showing the areas of coastal Peru that in the historical past have been subjected to Modified Mercalli seismic intensities of VIII or greater. The areas designated as intensity VIII may be larger because of population bias.

measured at the coast in the vicinity of the earthquake source region, can define a tsunami magnitude, $M_t$. The functional form is $M_t = a \log H_r + b$ where $a$ and $b$ are constants that depend on the source region and recording station. $M_t$ can subsequently be calibrated against the seismic moment-derived magnitude $M_w$.

Peru's long history of destructive earthquakes along its coastal zone is the direct result of the progressive underthrusting of the Nazca plate, which is moving at a rate of about 8 cm per year beneath the South American plate. The written historical record demonstrates that much of this movement has been accommodated by earthquakes with inferred magnitudes greater than 8. Of notable recent interest is the magnitude 7.9 shock that struck Peru in the vicinity of Yungay on May 31, 1970. A great rock- and mudslide called a *huayco* was triggered. Fatalities totaled 70 000 and 140 000 people were injured.

The spatial–temporal relation of offshore Peru earthquakes (between latitudes of ~6° to 20° S is shown in Figure 8.2. The rupture zones for the earthquakes of Peru can clearly be separated into three zones: a northern zone from 6° to 10° S latitude, a central zone from 10° to about 14.5° S and a southern zone from 14.5° to 18.5° S. The northern zone was struck by a large event in 1619 and a damaging

Table 8.1. *Great South American subduction-zone earthquakes*

| Date | $L$ (km) | $H_r$ (m) | $M_w$ | $M_t$ |
|------|------|------|------|------|
| February 8, 1570 | | | | 8–8.5? |
| December 16, 1575 | | | | 8.5? |
| January 22, 1582 | 80 | 1–2 | 7.5 | 7.7–8.0 |
| July 9, 1586 | 175 | 5 | 8.1 | 8.5 |
| November 24, 1604 | 450 | 10–15 | 8.7 | 8.8–9.0 |
| February 14, 1619 | 100–150 | | 7.7–8.0 | |
| May 22, 1664 | 75 | | 7.5 | |
| June 16, 1678 | 100–150 | 5 | 7.7–8.0 | 8.5 |
| October 20, 1687 | 300 | 5–10 | 8.4 | 8.5–8.8 |
| October 21, 1687 | 150 | | 8.0 | |
| August 22, 1715 | 75 | | 7.5 | |
| January 25, 1725 | 75 | | 7.5 | |
| December 24, 1737 | | | | 7.5–8.0? |
| October 28, 1746 | 350 | 15–20 | 8.6 | 9.0–9.2 |
| May 25, 1751 | | | | 8.5? |
| May 13, 1784 | 300 | 2–4 | 8.4 | 8.0–8.4 |
| September 18, 1833 | 50–100 | | 7.2–7.7 | |
| February 20, 1835 | | | | 8? |
| November 07, 1837 | | | | 9.3 |
| August 13, 1868 | 500 | 14 | 8.8 | 8.9 |
| May 10, 1877 | | 21 | | 9.0 |
| January 31, 1906 | 520 | | 8.8 | |
| August 17, 1906 | 330 | | 8.2 | 8.4 |
| November 11, 1922 | 330–450 | 9 | 8.5 | 8.7 |
| December 1, 1928 | 90 | | 7.6 | |
| May 24, 1940 | 220 | | 8.2 | |
| May 14, 1942 | 110 | | 7.6 | |
| August 24, 1942 | 200 | | 8.2 | |
| April 06, 1943 | 150–250 | | 8.2 | 8.2 |
| January 19, 1958 | 110 | | 7.7 | |
| May 22, 1960 | 1000 | 25 | 9.5 | 9.4 |
| October 17, 1966 | 120 | | 8.1 | |
| May 31, 1970 | 130 | | 7.9 | |
| October 03, 1974 | 180 | | 8.1 | |
| December 12, 1979 | 200 | | 8.2 | |
| November 12, 1996 | 100 | | 7.9 | |

Least-squares fit to all data. $M_w = 1.59 \log L + 4.49$; $M_t = 1.1 \log H_r + 7.7$, where $M_t$ is tsunami magnitude and $H_r$ is local tsunami height in meters (Abe, 1979; Dorbath *et al.*, 1990). $L$ is inferred from length of aftershock zone or length of region enclosed by Modified Mercalli intensity VIII contour.

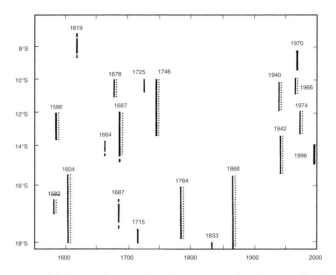

**Figure 8.2** Rupture length of offshore Peru earthquakes as a function of time. The variability in the scale of latitude is because length is measured offshore roughly parallel to the coastline and not along a meridian. The added dotted lines indicate earthquakes that triggered a tsunami (updated and revised from Dorbath *et al.*, 1990).

event in 1970. The combined seismic slip for these two events is far less than would be expected from the Nazca plate convergence rate of 8 to 10 cm per year, arguing that aseismic slip must also be taking place in this northern segment. The space–time distribution of great earthquakes shows that the central zone has the highest number of events over a 400-year period but also that it has two centuries of quiescence following the 1746 event.

One can estimate the total seismic slip in the central Peru zone since 1586 using an empirical scaling law proposed for subduction-zone earthquakes. In this correlation $d = 2 \times 10^{-5}L$, where $d$ is the slip, in meters, and $L$ is the rupture length, in meters, assigned to each earthquake. From 1586 to 1996 the total computed seismic slip was 37 m. This translates to a slip rate of about 9 cm per year, comparable to the total rate of plate convergence. Thus, even though the individual earthquake occurrences in this region are not equally spaced in time their overall rate of occurrence is in good agreement with the average rate of plate movement. The southern region has experienced four great earthquakes, the first in 1604, with varying lengths of rupture. Cumulatively the total seismic slip over the past four centuries is in good agreement with the total amount of convergence.

Dorbath *et al.* (1990) stated that "the time elapsed since the last 1868 event is now 120 years, a value which indicates that a great earthquake is highly probable [in the southern zone] in the very near future." A magnitude 8.4 earthquake struck Peru's southern region on June 23, 2001. This earthquake, the direct result of reverse faulting at the interface of the Nazca plate moving beneath the South American plate, validates the concept that earthquakes tend to occur in those regions of a plate boundary, a seismic gap, that have not experienced an earthquake on a time scale of hundreds of years.

## 8.2    Evidence for prehistoric events

But what of earthquakes of earlier times and events in the interior highlands? Certainly the continuing impingement of two colliding plates must have had some effect on the ebb and flow of prehistory in the Andean environment. Periodic natural disasters, such as those produced by onshore and offshore earthquakes and the climatic extremes of El Niño and La Niña, undoubtedly combined to produce a high level of environmental stress, at times leaving a recognizable record. Evidence for ancient earthquakes has been found in dry caves examined in the Ayachuco Basin of highland Peru near the Mantaro River (13.2° S, 74.2° W, Figure 8.1). Remains of early civilizations have been stratigraphically exposed in finely sliced vertical profiles at Pikimachay ("flea cave"). Pikimachay was formed by volcanic activity during the Quaternary time period and had ample remains for radiocarbon dating together with a long chronological stratigraphic sequence. Its dry environment was ideal for the preservation of plant and other remains. Substantial rockfalls, induced by violent shaking from one or more major earthquakes, occurred from 9000 BC to 7000 BC and from AD 300 to 900. The earlier stratigraphic horizon overlies early stone tools and sloth bones that have been dated with a radiocarbon date of ~12 000 BC found in a reddish orange soil of the excavation (Figure 8.3). It is clear that earthquakes must also have played a role in the evolution of early Andean cultures.

The physical world of coastal Peru is one of the most geologically active on Earth. It is continually subjected to earthquakes, tsunami, and the alternating flooding and drought effects of El Niño. There is evidence, for example, that the environment of the north coast of Peru experienced extended climatic disruption during the sixth century AD. The climate was dominated by droughts and severe flooding. Evidence for this disruption comes from extensive analyses of cores from ice caps and glaciers in the Peruvian highlands.

What about earthquake legends? It is highly improbable that early coastal communities did not experience earthquakes. As a result earthquakes may also have been exploited as a manipulative tool of the priestly leaders. Pachacámac, near Lima, was the center of an important ceremonial network on the central coast of Peru for many hundreds of years preceding the Spanish invasion of

STRATIGRAPHIC SEQUENCE IN SOUTH ROOM
AT PIKIMACHAY

| Matrix Description | | Radiocarbon Determinations |
|---|---|---|
| Dung | | AD 900—1200 |
| Charcoal | | AD 650—1250 |
| Ash | | AD 700—1100 |
| Rock and gray ash | | AD 550—950 |
| Rockfall | EARTHQUAKE | AD 300—900 |
| Charcoal | | 5650—5050 BC |
| Dark-gray brown ash | | 5800—5200 BC |
| Dark-gray ash | | 7200—6600 BC |
| Rockfall | EARTHQUAKE | 9000—7000 BC |
| Orange soil | | 11000—9000 BC |
| Reddish-orange soil ◄ | | 12380—12120 BC |
| Yellowish-brown soil | | 13000—12380 BC |
| Dark-brown soil | | 14150—11350 BC |
| Reddish-brown soil | | 15300—12900 BC |
| Dark-reddish-gray soil | | 19300—17200 BC |
| Brownish-gray soil | | 23000—19000 BC |

Evidence of human occupation and
sloth bones found here

**Figure 8.3** Stratigraphic sequence found in the South Room of Pikimachay. Two rockfalls were found in the sequence. Evidence of human occupation and sloth bones were discovered (data from MacNeish *et al.*, 1981).

the Inca empire. It is well known as the Inca site of the Pyramids with Ramps. *Pachacámac* was a feared deity of the Incas and one of his attributes was the control of earthquakes. The priests could use the threat of an earthquake to extract tribute, encourage political and economic collaboration, and punish offenders.

The reliance on adobe construction, and subsequent weathering by the elements, makes inferences of past earthquake occurrences subtle and difficult. Let us look at an example. Chimú was an important kingdom on the northern Peruvian coast which flourished from AD 1200 to 1450. Its capital was Chan-Chan located on a sandy low rise northwest of the Moche Valley flood plain (Figure 8.1). Chan-Chan was not occupied at the time that Europeans arrived in Peru. After abandonment, squatters broke holes in walls and irrigation channels. The site was subsequently destroyed by miners, looters, and later settlers, mainly in the search for gold and silver. Some archaeological investigators have argued that because the ancient adobe walls at Chan-Chan had joints, this method of construction was an adaptation to the dangers of earthquakes. However, this seems unlikely. When the very strong earthquake of May 31, 1970, struck the coastal region near Chan-Chan it was observed that modern adobe walls collapsed whether or not they had joints. Any remaining very thick walls at Chan-Chan withstood the earthquake and its aftershocks without damage. It is the width-to-height ratio of an adobe wall, and not the presence of joints, that gives stability during earthquake shaking. Any thin, tall adobe walls that might have been constructed in the historical past at Chan-Chan probably collapsed during earthquakes.

Early civilizations in the coastal valleys of the north coast of Peru developed an extensive irrigation system to develop more land for agriculture. In several locales the terrain allowed for the construction of inter-valley canals. This multi-valley irrigation linkage allowed for the development, maintenance, and sustenance of large population centers linked by a horizontal economy. The extensive courses of irrigation canals and aqueducts would, however, have been vulnerable to any elevation changes produced by tectonic uplift, subsidence, or geological fault movement. Any ground slope change would affect the distribution of water in a coastal drainage system. Synchronous tectonic land movement over wide regions has been observed and documented for many South American coastal earthquakes. The cumulative effects of elevation changes from earthquakes and any continuing aseismic movement would have had a profound effect on any natural or man-made drainage complex. Such evidence, in the form of reverse grades in canals, has been found. It is believed that these elevation changes produced far-reaching effects on agriculture and contributed to agrarian collapse.

About 100 km south of Chan-Chan, the Río Santa enters the sea slightly north of the coastal town of Chimbote. The Santa Valley was the locus of a pre-Hispanic coastal Peruvian society from AD 700 to 1100. This region contains the archaeological remains of habitation sites, pyramidal mounds, canals, rock-lined roads, and extensive sections of walls with fortresses and citadels. One of the significant engineering feats of an extensive nature in Peru was the construction of a "Great Wall System" on the north side of the Río Santa extending some 60 km eastward into the Andean foothills (Figure 8.4). The wall system consists of four to five major continuous sections with some intermittent gaps. In general the height of the wall is 2 to 2.5 m. The walls are of a mixed configuration and construction and include solid adobe blocks, dry angular solid rock, and various combinations thereof. Some sections of the wall, constructed of dry-laid angular rocks have a base width of 1.5 m and a battered face whereas other sections have a rectangular shape with a width of 0.75 m. It is believed that the wall was constructed in pre-Chimú times around AD 900. It has been suggested that the wall was constructed for defensive purposes. This theory is based on the presence of small forts, some circular and some rectangular, at irregular intervals on both sides of the wall but difficult to recognize from the valley floor below (Figures 8.5 and 8.6). Other investigators have argued that because of its discontinuous segments the wall was not a defensive barrier but instead a symbolic

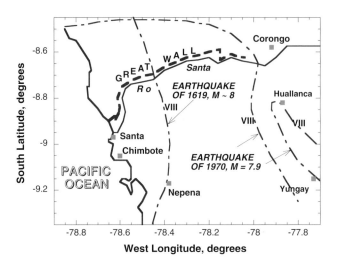

**Figure 8.4** Map showing approximate location of the pre-Chimú great wall on the north coast of Peru. Areas enclosed by isoseismals of Mercalli intensity VIII or greater are shown for the Peruvian earthquakes of 1619 and 1970. It is most likely that the wall was subjected to damaging seismic intensities in the historical past.

barrier for either social, ethnic, or traffic and commerce purposes. Fallen sections of the wall exist and clearly some sections of the wall have been repaired, suggesting damage from past earthquakes (Figures 8.7 and 8.8). The wall does not cross any mapped surface faults and lies in a zone of Tertiary volcanic rocks. However, the wall has been subjected to two $M \sim 8$ earthquakes, in 1619 and 1970, and there is no reason, considering the tectonic environment, that other substantial-sized earthquakes could not have pre-dated these occurrences. The isoseismal areas encompassing a Mercalli intensity of VIII or greater from these events are shown in Figure 8.4. Intensities of VIII or greater are associated with levels of ground accelerations approaching 20% $g$ or more and would be severe enough to produce damage to dry-laid stone walls.

## 8.3    Earthquakes at Cuzco

Cuzco, an inland city with legendary prestige in southern Peru, was founded in the eleventh century by Manco Capac, the first of the Incas. Architecturally it is a strange conglomeration of three distinct building types: Inca stone walls, and Spanish colonial and adobe buildings. Some houses and buildings are built on the foundation of massive Inca stone work. The church and convent of Santo Domingo incorporates in its walls part of the Inca "Temple of the Sun" (Corichana). Cuzco's cathedral is a famous example of early Spanish colonial architecture. Crude adobe buildings are still in use today.

Figure 8.5 Aerial photograph of the "Great Wall" along the crest of a spur of the Andes on the north border of the Santa Valley (Shippee, 1932; © 1932, American Geographical Society).

Figure 8.6 One of the square fortresses or citadels along the wall on the Santa River (Shippee, 1932; © 1932, American Geographical Society).

**Figure 8.7** A fallen section of the Santa River Wall (Shippee, 1932; © 1932, American Geographical Society).

**Figure 8.8** A section of the wall showing the angular rock construction and the clear zone of smaller stones indicating later repair (Shippee, 1932; © 1932, American Geographical Society).

Cuzco has been subjected to a number of damaging earthquakes, some with interesting religious ramifications. The most notable earthquake took place on March 31, 1650 with a lasting effect on the earthquake lore and history of the town. Any estimate as to the size of this event would only be a guess but it did destroy nearly all of Cuzco and was felt at Lima. Only a few important buildings were left standing. These were Cuzco's famous cathedral; the church of Santa Clara; the site of its oldest convent; and Santo Domingo church. The 1650 earthquake was the most disastrous event in three centuries of Spanish rule. In the *Noticias cronologias de la gran ciudad del Cuzco* it is written by priests that the earthquake lasted the time for the populace to say "two or three creeds." More secular individuals remarked that the quake lasted about 15 minutes. This stated duration must have been greatly exaggerated but it is a typical time warp given by individuals even today when describing the length of time that they experience shaking from large earthquakes.

Descriptions stated that individuals ran in confusion without knowing where to go and that men in their panic did not look out for their wives, women, or children. The Indians of Cuzco thought themselves doomed and in their attempt to appease the "wrath of God" carried an icon of the crucified Christ around the main plaza. This is now an annual act of devotion. In Cuzco's cathedral is the Capilla del Señor de Los Temblores with a replica of the famous Spanish statue Cristo de Cachorro; the Indians, having been spared further havoc, called the statue "Lord of the Earthquakes." Between March 31, 1650 and January 1, 1651, the total number of felt temblors and aftershocks is given as 823. The various subtotals given by the individual priestly compilers do not tally correctly, emphasizing either exaggeration or bad arithmetic. The disaster of 1650 did produce a significant drop in the indigenous population, who abandoned their homes and migrated from the city. Most of these individuals lived in earthquake-vulnerable dwellings and structures. On September 17, 1707, another strong earthquake shook Cuzco and several outlying towns, producing 160 fatalities and many injuries. Aftershocks continued until October 7. Of historical interest is the fact that the quake produced another kind of exodus. Many Indians with their shamans convened in the epicentral area to practice superstitious rites in the hope of preventing further aftershocks. It was clear to the ecclesiastics that *el execrable crimen de la idolatría* ("the abhorrent crime of idol worship") would perpetuate heresy and must be dealt with by a saintly tribunal. The penalty given to an unfortunate few was 200 lashes.

Other earthquakes of interest to strike the Cuzco region were the $M = 6$ event of September 18, 1941, and the $M = 5.3$ event of April 5, 1986. The 1941 event killed 129 people and required rebuilding of 63 percent of the dwellings of Cuzco. Much of the damage was confined to Spanish colonial churches, particularly the bell towers, and old adobe houses which lacked wall reinforcement and adequate ties of the upper floor and roof to the walls. Ancient stone construction of the Incas showed no major effects from the earthquake. The $M = 5.3$ event produced more surprising seismological results. There were a small number of fatalities and some damage in Cuzco but this moderate-magnitude earthquake did produce clear surface ruptures in the epicentral area some 10 km to the north. The area of maximum Mercalli intensity of VIII was centered along the zone of largest surface offset. These ruptures totaled 3 km in length with segments exhibiting up to 10 cm of vertical offset or throw. Normally surface ruptures or faulting are only associated with earthquakes of $M = 6$ or more.

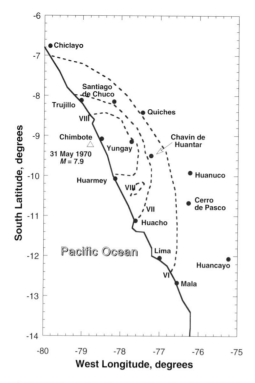

Figure 8.9 Isoseismal map of the May 31, 1970 Peru earthquake (after Gajardo, 1970).

## 8.4    Disastrous landslides

One of the most catastrophic historic earthquakes in the Americas struck Peru on May 31, 1970. Even though this $M = 7.9$ earthquake had its epicenter offshore, 25 km west of Chimbote, the affected onshore area was 65 000 km$^2$ (Figure 8.9). The pattern of aftershocks covered an area paralleling the coast 140 km long by 50 km wide. Fatalities totaled 70 000, with an additional 140 000 people injured; 80 percent of the structures in the stricken area were rendered uninhabitable. The earthquake was so devastating because it triggered a cataclysmic debris avalanche that inundated the inhabitants of Yungay, and 18 000 people were buried alive. A large mass of rock capped with ice and snow moved downhill from its starting point at a height of about 6000 m at Huascarán Mountain, 14.5 km east of Yungay (Figure 8.10). The highly fluid debris flow with a volume of 50 to 100 million m$^3$ moved downhill with very high velocity, estimated at 280 to 335 km/h, oversweeping a 100- to 200-m-high ridge and burying most of the city with 4 to 5 m of mud and rocks.

An eyewitness account by a geophysicist M. Casaverde of Peru describes the horror of the Huascarán debris avalanche that struck Yungay:

**Figure 8.10** View looking east showing the 1970 Huascarán debris avalanche (photo courtesy of National Geophysical Data Center).

I was in Yungay . . . on a tour of this area . . . I realized we were experiencing an earthquake . . . when the shaking began to subside . . . I heard a great roar coming from Huascarán . . . I ran . . . toward Cemetery Hill . . . when I felt a strong turbulent blast of wind . . . I reached the upper level of the cemetery . . . as the debris flow struck the base of the hill . . . I remember two women who were no more than a few meters behind me and I never did see them again . . . I counted 92 persons who had saved themselves . . . mostly teenagers . . . the older could not run fast enough to escape. It was the most horrible thing I have ever experienced and I will never forget it. (quoted in Cluff, 1971)

An additional 20 000 fatalities were the direct result of the failure of adobe and masonry structures which contain little resistance to the lateral forces imposed by seismic shaking. The area of moderate to heavy building damage and ground

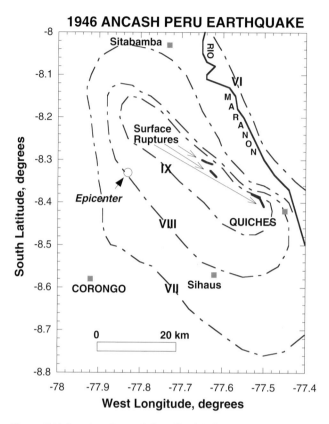

**Figure 8.11** Isoseismal map (after Silgado), for the Ancash, Peru earthquake of November 10, 1946. The zone of visible ruptures is shown together with the instrumentally determined epicenter. The isoseismal VIII bounds approximately 2300 km². Isoseismal IX encloses an area of 1600 km². In view of the surface rupture, intensities of X must have been reached within that area.

shaking occurred along the coast from Trujillo in the north to Huacho in the south. Soil liquefaction, fissuring, and surficial subsidence was particularly severe in Chimbote. A maximum Modified Mercalli intensity of VIII was assigned to a portion of the coastal area and in an elongated valley area inland from the coast. Ground shaking was reported to have been primarily a strong side-to-side motion that lasted 30 to 45 s with adobe structures beginning to collapse after 15 s or so of hard shaking. The earliest destructive earthquake to affect the region occurred in 1619. This shock was especially destructive at Trujillo on the coast and was reported to have generated landslides in the interior. Another earthquake damaged Trujillo on January 6, 1725, triggering an ice and mud avalanche that killed 1500 people and buried the town of Ancash, near Yungay. This sector of the Peruvian coast had not been subjected to major destructive

Figure 8.12 The 3.5 m-high fault scarp near Quiches, formed during the Ancash, Peru earthquake of November 10, 1946. Downthrown side toward the west (from Silgado, 1951).

earthquake activity since 1725. It is well documented, however, that numerous non-seismically triggered landslides have struck the region in the past.

Considering the size of the event and the amount of devastation, the earthquake was notable for its lack of other concurrent geological or tectonic effects. There was an absence of any movement, sympathetic or otherwise, on known surface faults within the area. Additionally, no tectonic warping took place along the coastal region. Finally, there was no tsunami or seismic sea wave generated. The causative fault must have ruptured at depths well beneath the ocean floor. This devastating event brought strange words into the local usage of Spanish: *damnificado* (one who sustained material losses as a consequence of the earthquake) and *terremotear* (a made-up verb meaning "to earthquake").

On November 10, 1946, a magnitude 7.2 event known as the Ancash (a political unit), Peru, earthquake struck the highlands of the Andes almost completely demolishing the town of Quiches (Figure 8.11). The earthquake was extremely damaging, with 1400 fatalities, quite surprising considering the remoteness of the epicentral area. The faulting observed broke the Earth's surface discontinuously over a length of at least 20 km; vertical scarps reached a maximum height of 3.5 m (Figure 8.12). The area outlining an apparent intensity of IX covered approximately 1600 km$^2$. There are probably intensities of X within this area. As is typical of many earthquakes where damage is more noted in scattered districts of population, the instrumentally located epicenter is not positioned within the zone of maximum intensities. The earthquake is notable because it was the first South American earthquake where accompanying major surface faulting was well documented. Because the fault rupture took place in a remote,

**Figure 8.13** Map showing the location of the Mantaro River landslide of April, 1974.

sparsely populated region, damage by direct fault displacement was not a major factor. One of the main reasons for its destructiveness, besides the intensity of its shaking, was the large number of landslides triggered; these landslides were responsible for 500 fatalities and one of the stricken villages was buried to a depth of 20 m.

Rockslides and landslides are often triggered by earthquakes. However, it is unusual when such sudden earth movements are initiated by processes other than earthquakes and also generate measurable seismic signals. On April 25, 1974, a large rockslide occurred along the Mantaro River in Peru (Figure 8.13). The total volume of the rockslide/debris flow produced by a failure on a geological bedding plane was about $1.6 \times 10^9$ m$^3$; 450 people lost their lives. A portion of the gravitational energy lost in the rapid downward movement of the rock mass was converted to seismic waves. The recorded seismic event produced a signal equivalent to that from an $M = 4.5$ earthquake. Assuming that the maximum seismic-wave amplitude (and hence the determined magnitude) was produced by the initial rockfall, which fell 500 m or so with a volume equal to 50 percent of the total rockslide/debris flow, the gravitational potential energy release is estimated at $5 \times 10^{22}$ ergs ($10^7$ ergs = 1 joule). A magnitude 4.5 earthquake indicates a radiated seismic energy release ($\log E = 11.8 + 1.5\,M$) of $3.5 \times 10^{18}$ ergs, therefore the conversion, or efficiency of transfer, from gravitational potential energy to radiated seismic energy release is low. A similar conclusion of low conversion efficiency was also reached from analyses of the seismic signals produced by a fatal rockfall at Yosemite National Park, California (Uhrhammer, 1996).

## 8.5    Tsunami of the coast of Chile

Chile is a long, narrow, ribbon-like country on the west coast of South America, occupying a latitudinal range from 17° 25′ to 55° 59′ S. Bounded by Peru on the north, Chile is climatically and geographically divided into three regions. North of 30° S is the mineralized desert region where rain rarely falls. The principal cities are Arica, on the border between Chile and Peru; Iquique; and Antofagasta near the Tropic of Capricorn. Copiapó and Coquimbo are at a latitude of 30° S. From 30° to 40° S is the verdant central zone containing the main cities of Valparaíso, Santiago, and Concepción. The main city in the forested, southern rainy zone south of 40° is Valdivia (Figure 8.14). All of Chile is subjected to the effects of earthquakes. The coastal region of Chile is particularly vulnerable to the effects of tsunami or seismic sea waves and often acts as their geographical location.

Tsunami are initiated by a vertical movement of the seafloor and less frequently by submarine landslides or volcanic activity. Because water is incompressible, a sudden mound or trough of water forms at the ocean surface above the disturbance. Gravity immediately acts to level the surface. The sudden movement of water to restore a level situation sets up an oscillating wave that travels outward at speeds of several hundreds of kilometers per hour. In deep water the amplitude or height of the wave is only a few meters but its wavelength or distance from crest to crest is hundreds of kilometers. As a result the passage of a tsunami is not noticed by a ship in the open seas. When a tsunami wave approaches shallow water it undergoes a change. The energy in a column of water is proportional to the square of its height times its wavelength, $H^2L$. Approaching the shore there will be a decrease in wave speed due to shallowing. Since the wave frequency is constant the wavelength of the water wave decreases but conservation of energy requires an increase in wave height. The

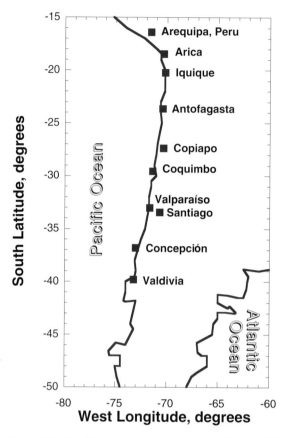

**Figure 8.14** Location map showing the main cities of Chile.

advancing wave front steepens and increases in height until the wave breaks. A tsunami can be particularly devastating in funnel-shaped bays or estuaries with broad mouths.

Several tsunami that have struck the Chilean coast are of historical interest. On August 13, 1868, a great earthquake occurred off the coast of southern Peru. Estimates of its magnitude range from 8.5 to 8.8. The towns of Arequipa in Peru and Arica in Chile, were largely destroyed. The resulting tsunami damaged waterfront property at distant locations in New Zealand, Australia, Samoa, Hawaii, San Pedro (California), and Japan. At Arica the sea withdrew about 20 minutes after the onset of the earthquake. The first wave to reach Arica was ~10 m above the high water mark, followed by subsequent waves, the third and fourth reaching a height of 14 m. Of note is the fact that a visiting ship the USS Wateree was carried intact 5 km from its anchorage and was placed 3 km inland (Figure 8.15). Its naval crew was involved in land rescue operations! The USS

**Figure 8.15** In Arica, Chile, on August 13, 1868, the sea withdrew about 20 minutes after an $M = 8.5(?)$ earthquake struck. Subsequent tsunami waves with heights reaching 14 m carried several ships several miles inland. The USS Wateree shown in the foreground remained intact and its crew were involved in land rescue operations. (Photograph courtesy of Branner Library, Stanford University.)

**Figure 8.16** View of the beach area at Arica, Chile, after the $M = 8(?)$ earthquake and tsunami of May 9, 1877. Note the railroad car on the beach. The USS Wateree, which had been beached inland by the 1868 tsunami, was refloated, coming to rest near the shore several kilometers to the north of its initial position. Observe the height of wave cutting on the cliff in the background. (Photograph courtesy of Branner Library, Stanford University.)

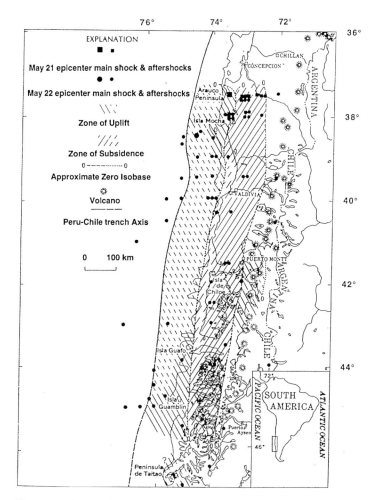

**Figure 8.17** Zones of land level change and earthquake epicenters associated with the 1960 Chile earthquake and their relation to the axis of the Peru–Chile Trench and the Andean volcanic chain (after Plafker and Savage, 1970).

Wateree would have remained grounded at its location if it had not been for a subsequent earthquake. In the early evening of May 9, 1877, a great earthquake ($M = 8$ to 8.5) occurred offshore of Chile somewhere between Iquique and Antofagasta. The distant tsunami caused destruction in New Zealand, Acapulco (Mexico), Gaviota Beach (California), Hawaii, Samoa, and Japan. Locally the tsunami began about 5 minutes after the earthquake with estimates of the initial wave height ranging from 9 to 14 m. The second wave was stated to have reached 21 m in height. In Arica the shipwrecked USS Wateree was refloated by the tsunami waves and moved closer to shore, several kilometers north of its initial location. Damage to the beach area of Arica was severe (Figure 8.16).

Table 8.2. *Major earthquakes of Chile, 1562–1995[a]*

| Date | Locale | Magnitude | Observations |
|---|---|---|---|
| October 28, 1562 | Southern Chile | | Doubtful |
| February 8, 1570 | Concepción | 8.0–8.5 | Tsunami |
| March 17, 1575 | Santiago | 7.0–7.5 | |
| December 16, 1575 | Valdivia | 8.5 | Tsunami |
| November 24, 1604 | Arica | 8.5 | Tsunami |
| September 16, 1615 | Arica | 7.5 | Small tsunami |
| May 13, 1647 | Santiago | 8.5 | Onshore |
| March 15, 1657 | Concepción | 8.0 | Tsunami |
| March 10, 1681 | Arica | 7.5 | No tsunami |
| July 12, 1687 | Near Santiago | 7.5 | 1688? |
| July 8, 1730 | Valparaíso | 8.8 | Tsunami |
| December 24, 1737 | Valdivia | 7.5–8.0 | Tsunami? |
| May 25, 1751 | Concepción | 8.5 | Tsunami |
| March 30, 1796 | Copiapó | 7.5–8.0 | No tsunami |
| April 3, 1819 | Copiapó | 8.3–8.5 | 3 Events; tsunami |
| November 19, 1822 | Valparaíso | 8.5 | Tsunami |
| September 26, 1829 | Valparaíso | 7.0 | No tsunami |
| February 20, 1835 | Concepción | 8.0–8.3 | Tsunami |
| November 7, 1837 | Valdivia | 8.0 | Tsunami |
| October 8, 1847 | Illapel | 7.0–7.5 | |
| December 17, 1849 | Coquimbo | 7.5 | Tsunami |
| December 6, 1850 | Maipo Valley | 7.0–7.5 | Rockslides |
| April 2, 1851 | Casablanca | 7.0–7.5 | No tsunami |
| October 5, 1859 | Copiapó | 7.5–7.8 | Tsunami |
| August 13, 1868 | Arica | 8.5 | Tsunami |
| August 24, 1869 | Pisagua | 7.0–7.8 | Moderate tsunami |
| October 5, 1871 | Iquique | 7.0–7.5 | |
| May 9, 1877 | Pisagua | 8.0–8.5 | Tsunami |
| February 2, 1879 | Magellan | 7.0–7.5 | |
| August 15, 1880 | Illapel | 7.5–8.0 | No tsunami |
| August 17, 1906 | Valparaíso | 8.3 | Tsunami |
| December 4, 1918 | Copiapó | 7.5 | Moderate tsunami |
| November 11, 1922 | Huasco | 8.4 | Tsunami |
| December 1, 1928 | Talca | 8.4 | Small tsunami |
| January 24, 1939 | Chillán | 8.3 | No tsunami |
| April 6, 1943 | Illapel | 8.3 | Minor tsunami |
| December 17, 1949 | Punta Arenas | 7.5 | Minor tsunami |
| May 6, 1953 | Chillán | 7.5 | No tsunami |
| April 19, 1955 | North of Valparaíso | 7.0 | Tsunami |
| May 22, 1960 | South of Concepción | 8.5 | Tsunami |

(cont.)

Table 8.2. (*cont.*)

| Date | Locale | Magnitude | Observations |
|------|--------|-----------|--------------|
| February 14, 1962 | Concepción | 7.3 | |
| July 9, 1971 | North of Valparaíso | 7.8 | |
| May 5, 1973 | Valparaíso | 7.5 | |
| October 5, 1975 | Concepción | 7.8 | |
| November 30, 1976 | Northern Chile | 7.3 | |
| October 16, 1981 | Off Valparaíso | 7.2 | |
| November 7, 1981 | Northeast of Valparaíso | 6.5 | |
| October 4, 1983 | Copiapó | 7.3 | Minor tsunami |
| March 3, 1985 | Valparaíso | 7.8 | Small tsunami |
| April 9, 1985 | Santiago | 7.5 | |
| August 8, 1987 | Arica | 6.9 | Landslides |
| January 19, 1988 | South of Antofagasta | 6.7 | |
| February 5, 1988 | South of Antofagasta | 6.8 | |
| October 28, 1992 | North of Valparaíso | 6.5 | |
| July 11, 1993 | Northern Chile | 6.2 | Landslides |
| July 30, 1995 | Antofagasta | 8.0 | |

[a]Updated from Lomnitz (1970a) using data from the International Seismological Centre.

Some insight into the mechanism of tsunami generation can be obtained by examining one of the major seismic events of the twentieth century. On May 21, 1960, a $M_w = 9.5$ earthquake struck the coastal region of Chile. This earthquake is the largest seismic event ever recorded. Tectonic movements, both uplift and subsidence, took place over a vast area relative to sea level. The length of the zone of rupture was about 1250 km. Uplift occurred offshore and subsidence along a coastal zone (Figure 8.17). Analyses of the tsunami waves at distant Pacific-rim stations are compatible with the observations that significant uplift of the ocean floor took place. The tsunami was recorded as an initial rise on tide-gage stations indicating that at the source there was a push to the water mass from an upward motion of the sea bottom. Tsunami run-up distances ranging from 10 to 20 m between Isla Guafo and Isla Mocha confirm large seafloor displacements at the source region. Significant destruction from the tsunami occurred in Japan, Hawaii, and the Philippine Islands.

Vertical movements associated with earthquakes are characteristic of this segment of the South American coast. An examination of the earthquake record in Table 8.2 shows that tsunami have accompanied about 50 percent of the major earthquakes recorded since 1562. Evidence for earlier seismic events is undoubtedly present in the recent geological record. Parts of the mainland coast of Chile and its offshore islands, such as Isla Chiloé, have been known since Darwin's visit in 1835 to exhibit marine terraces that are significantly uplifted above present sea level. Since these older terraces were located in areas subjected to *both* uplift and subsidence during the 1960 earthquake, a net rise or emergence of these regions is implied. Determination of the absolute ages of these terraces would indicate the past timing of any tectonic movements. The problem is not completely straightforward, however, because any determinations would be complicated by any glacial–eustatic changes in sea level.

Tectonic activity associated with the eastward movement of the Nazca plate has produced a long history of deformation and earthquakes along the western regions of Ecuador, Peru, and the northern part of Chile. These periodic transients, which have long affected the Andean environment, assuredly contributed to the ebb and flow of important cultural centers and initiated architectural and environmental adaptations to such natural disasters; these adaptations have yet to be uncovered in the archaeological record.

# Early California earthquakes

## 9.1　Los Angeles Basin earthquake of July 28, 1769

California is interlaced with hundreds of faults and no part of the state has been immune from felt earthquakes. Not all of the mapped faults have been responsible for major earthquakes and most of the significant historical events have usually been associated with the San Andreas fault system which is aligned in a northwesterly–southeasterly direction through the state. The written history of California earthquakes starts in 1769 with the expedition of Gaspar de Portola.

In July, 1769 Gaspar de Portola, governor of the Californias, and his followers, who included Miguel Costansó and Father Juan Crespi, set out from San Diego on a landward journey to Monterey. Each of these three travelers kept a daily diary of their experiences. In their northward trek the party halted along the banks of the Santa Ana River, near a place now known as Olive, 4 km east of the present-day city of Anaheim. On July 28, 1769, the travelers experienced four violent shocks beginning at 1 p.m.; the last shock occurred around 4 p.m. The first earthquake was described as lasting half as long as repeating an Ave Maria (the Hail Mary of the Roman Catholic Church, about 15 seconds to recite). Reportedly the earthquake shock frightened the Indians into a prayer to the four winds. Father Crespi named the river where they camped Jesús de los Temblores.

On the following day the party had moved northwesterly past Brea Canyon, crossing the Puente Hills near La Habra and making camp at the San Gabriel River, near Bassett. On the afternoon of July 30 another earthquake was felt. The party next camped at Lexington Wash, about 6 km east-southeast of San Gabriel. An earthquake was felt on the morning of July 31 and three shocks were felt on August 1 while the travelers were at rest. On August 2 the expedition moved westward stopping near the Los Angeles River where three consecutive earthquakes were felt during the afternoon and the night. On August 3 the party went westward, pitching camp at Ballona Creek, 5 km west of Cienega. Springs of bitumen and asphalt marshes were noted to the right (north) of their path (in the vicinity of the La Brea tar pits) leading to speculations that there might be a volcanic association with the earthquakes that were felt in the afternoon.

It is not possible from the descriptions given in the diaries to determine what causative fault in southern California was responsible for the main earthquake and its obvious aftershocks. Later writers have estimated the seismic intensities at VI to X on the Rossi–Forel Intensity Scale, corresponding to values of VI to VII to X+ on the Modified Mercalli Intensity Scale. The higher values seem improbable. The earthquake was *not* reported as felt by the astute Father Junípero Serra and two friars who remained behind in San Diego. Earthquakes were only reported as felt when the exploratory party was at rest or was camped. There were no descriptions of any of the horses or pack mules being startled. When earthquake shaking is felt *while moving* it is usually assigned a Mercalli intensity of VI to VII or greater. Given the circumstantial, descriptive evidence in the diaries and the 70 km distance traversed across the Los Angeles Basin it is believed that what was felt by members of the Portola expedition was a strong, *local* shock of magnitude 6 to 6+.

## 9.2    The earthquakes of 1812 and 1836

Written documentation of early California earthquakes can sometimes be found in the reports of the Franciscan Missions (Englehardt, 1897). The year 1812 is of note because it was referred to as *el año de los temblores*. Reports describing the events in 1812 are not particularly satisfactory but there were at least two destructive earthquakes in 1812, on December 8 and December 21. During the early hours of December 8, the day of the Feast of the Immaculate Conception, an earthquake in southern California affected many of the missions. At Mission San Juan Capistrano, about 55 km southeast of Los Angeles, the lofty tower at the church front (Figure 9.1) fell on the vaulted roof below (Figure 9.2). Forty people were killed and the church was never rebuilt. The building had thick walls and an arched, dome-like roof that was constructed of irregular-sized un-hewn stones set in mortar of poor quality. Statements were made in the written records that it was not the strength of the earthquake that caused the damage but rather the poor construction.

Local intensities of VII to VIII on the Mercalli Scale are inferred from the damage reported at the San Gabriel Mission about 20 km east of Los

**Figure 9.1** An artist's rendition of Mission San Juan Capistrano prior to the partial destruction by the earthquake of December 8, 1912 (Engelhardt, 1922).

Angeles. The main altar was overthrown and the top of the steeple was toppled. Damage was reported to the sacristy, the convent, and other buildings. It is not clear whether the earthquake was felt as far to the southeast as the missions at San Luis Rey or San Diego. No earthquake was reported on December 8 at Mission San Fernando, west of Los Angeles, but it was reported that the earthquake of December 21 *did no further damage* and only necessitated the introduction of 30 new beams to support the church wall. (This implies damage from an earlier quake.) At Mission La Purísima Concepción, 210 km northwest of Los Angeles, several shocks were felt in the early morning of December 8 but no harm nor damage was reported. (Some confusion in dating has crept into the written reports. The reported damage to Mission San Juan Capistrano on *Sunday*, December 8, actually took place on *Tuesday*, December 8, 1812.)

Unusual distress to trees was noted in 1988 along the San Andreas fault in the Wrightwood area, some 50 km northeast of San Gabriel (Figure 9.3). The trees studied showed suppressed ring growth beginning in 1813, probably caused by root severance during fault slip movement as a result of the earthquake of December 8, 1812. (The earthquake of December 21, discussed below, was more likely an offshore event positioned further to the northwest). The documented trauma to the trees suggests that the December 8 event was a San Andreas fault event and it is interesting to speculate on the size of this event. Excavations at Pallett Creek, about 50 km northwest from Wrightwood along the strike of the

**Figure 9.2** Destruction of the stone church at Mission San Juan Capistrano caused by the earthquake of December 8, 1812. The lofty tower at the church front in Figure 9.1 collapsed on the arched roof (Engelhardt, 1922).

San Andreas fault, found 6 m of right-lateral offset, probably generated from the December 8 event. This observation, together with the inferred level of shaking at the missions of San Fernando, San Gabriel, and San Juan Capistrano point to a fault rupture length of at least 50 km. This rupture length suggests a minimum moment magnitude of 7.1 for the December 8, 1812 event.

On the southern segment of the San Andreas fault attempts have been made to correlate individual historical seismic events, such as the 1812 event, by using radiocarbon dates of ancient earthquakes at adjacent sites. Such a study, using data from the nearby sites of Pallett Creek and Wrightwood, is summarized in Table 9.1.

Table 9.1. *Recalibrated $^{14}C$ dates for San Andreas seismic events*[a]

| Pallett Creek | Wrightwood |
| --- | --- |
| 1857 event | 1857 |
| 1753–1772 | 1812 event |
|  | 1682–1733 |
|  | 1566–1567 |
| 1473–1551 |  |
|  | 1454–1455 |
| 1352–1360 |  |
|  | 1160–1175 |
| 1046–1078 |  |
| 861–1014 |  |
| 893–950 |  |
| 814–835 |  |
| 714–758 |  |
| 519–637 |  |

[a] Biasi and Weldon (1994).

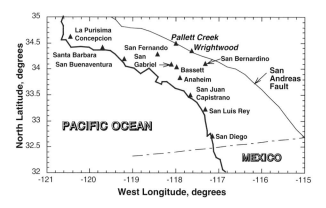

Figure 9.3 Map of southern California showing the locations of the missions discussed in the text. Wrightwood is the area where trees are believed to have been disturbed during the 1812 earthquake (adapted from Jacoby *et al.*, 1988).

The paleoseismic observations at Pallett Creek show nine individual events pre-dating the 1857 earthquake (see Section 9.4) with an average return time of 132 years. At Wrightwood only five events with a return time of 129 years are seen. However, there is a striking non-correlation of individual dated events arguing that the fault slip for these occurrences was highly irregular in both space and time. It cannot be overemphasized that paleoseismology on the San Andreas fault system is not an easy matter. Sites have to be selected where the sedimentary layers have been deposited in a steady, uninterrupted fashion and where the sites contain materials that allow the layers to be dated by radiocarbon methods.

The earthquake of December 21, 1812 appears to have produced more damage at the northern missions. At Santa Barbara several shocks were reported in 1812 (specific days not stated). The damage was described as severe in that all of the mission buildings were damaged, together with the church, which was not repaired since it was subsequently elected to rebuild a new one in its place. Two shocks occurring 15 minutes apart are described as striking Santa Inez, 40 km east-northeast of Santa Barbara, destroying a corner of the church, destroying 25 percent of the new houses in the proximity, ruining the mission roofs, and damaging many walls. At Mission San Buenaventura, on the California coast about 50 km southeast of Santa Barbara, damage to the church was so severe that the tower and much of its façade were eventually rebuilt. It was recorded that the site upon which the mission was built appeared to have settled and that inhabitants, fearful of being inundated by the sea, departed and did not return until the following year. Damage was severe at Mission Purísima Concepción, 70 km west northwest of Santa Barbara and within the present city limits of Lompoc, California. The earthquake(s) brought down the church, almost all of the mission buildings, and about 100 adobe houses. There were reports of ground fissures that emitted water and black sand. Subsequent flooding from heavy rains completed the damage and a new mission complex was built slightly north of the damaged site.

In assessing the size of this event it is important to emphasize that the reported damage pertains to the early adobe churches which are synonymous with weak masonry. A damaged church that needs some rebuilding correlates well with a Modified Mercalli intensity of VII. Knocking down a church and most of the convents and houses in the vicinity indicates an intensity of at least IX. The reported limit of damage encompassed an area bordered on the extremities by Purísima Concepción to the north and San Buenaventura to the south. From this a reasonable estimate of the area shaken at an intensity of VII or greater, excluding the area offshore, is about $12\,000$ km$^2$. This area would indicate a value of $M_w = 7.1$ as also being appropriate for the event of December 21, 1812. The epicenter was probably located offshore in the Santa Barbara Channel somewhere between Pt. Conception and Santa Barbara.

The California missions were secularized in 1834 and the writing of regular reports from the missions was suspended. As a result information about earthquakes in California, from that date until regional newspapers appeared during

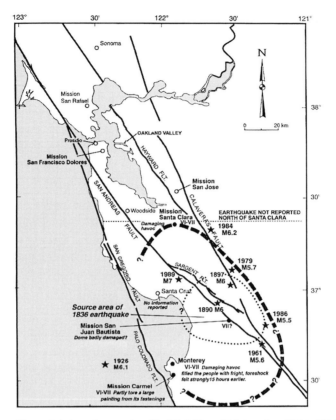

**Figure 9.4** Area damaged at Modified Mercalli intensity of VI to VII from the earthquake of June 10, 1836, shown by the dashed curve. The stars indicate the locations of eight recent events for which damage comparisons to the 1836 event were made (taken from Toppozada and Borchardt, 1998).

the 1849 Gold Rush, is incomplete and patchy. Examples of this patchy information are the reports pertaining to the earthquake of June 10, 1836, sometimes referred to as the "1836 Hayward fault earthquake." Careful re-evaluation of the limited data has produced the isoseismal map shown in Figure 9.4. Bearing in mind that the region enclosed by Mercalli intensity values of VI to VII is constrained solely by the observations at Carmel, Monterey, and Santa Clara, the contoured area of intensity VI to VII is approximately 5500 km$^2$. If the contour is taken to enclose an area shaken at an intensity of VII or greater, then empirical relations would suggest an $M \sim 6.7$. If intensity values in the range of VI to VII are assumed, a value closer to $M \sim 6.2$ is deduced. Regardless of the choice this event can at best only be described as a large event that took place in the general area between Santa Cruz and Monterey, but certainly not on the Hayward fault.

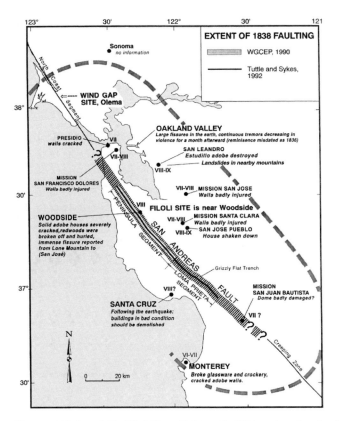

**Figure 9.5** Inferred Modified Mercalli intensity VII isoseismal and estimates of surface rupture for the 1838 earthquake. Locations of fault trenching sites at Olema and Filoli are shown (adapted from Toppozada and Borchardt, 1998). WGCEP, Working Group on California Earthquake Probabilities.

## 9.3    The 1838 San Andreas fault earthquake

A large earthquake(s) took place in June and July of 1838 (exact days unknown) with reported damage in San Francisco, Oakland, Monterey, and the missions from San Francisco, southward to San Juan Bautista. An inferred iso-seismal map together with the postulated extent of surface faulting is shown in Figure 9.5. From the area enclosed by the isoseismal contour for intensity VII (shown by the dashed curve) one would infer a magnitude of about 7.4 for the 1838 earthquake. Estimates of the 1838 fault rupture length, $L$, which include the Peninsula and Loma Prieta segments of the San Andreas fault zone, range from 100 to 140 km. Using the empirical relation between moment magnitude and rupture length (Wells and Coppersmith, 1994), $M_w = 5.08 + 1.16 \log L$, one obtains values of $M_w$ ranging from 7.4 to 7.6, in very good agreement with the estimate based on the isoseismal area of intensity VII.

The 1838 fault segment ruptured again during the 1906 San Francisco earth-quake. A portion of the 1838 segment failed again as a result of the 1989 Loma Prieta earthquake on the San Andreas fault. It is an interesting speculation whether 1906-sized events occur following large earthquakes on segments of the subsequent rupture. Fault trenching at the Filoli site near Woodside has allowed the matching and radiocarbon dating of several offset buried stream channel deposits across a portion of the San Andreas fault that caused surface rupture in 1906. One recent channel that pre-dates the June 1838 event by at least tens of years was offset 4.1 m, and includes the 1906 offset of 2.5 m, inde-pendently observed in the vicinity of Woodside. The additional offset of 1.6 m, which is compatible with a magnitude 7.1 event, is attributed to the 1838 event. By matching displaced or offset steam channels at Filoli, Hall *et al.* (1995) suggest a late-Holocene slip rate of 17 ± 4 mm per year for the San Andreas fault in this vicinity. If this rate is uniform, at least 1.4 m of potential offset has accumulated since 1906, pointing to the potential for another 1838-sized event ($M = 7.4$) in this vicinity.

### 9.4    Fort Tejon, California earthquake of January 9, 1857

The 1857 Fort Tejon earthquake is often pointed to as the largest event known in California's historical record. Despite its size, the earthquake caused little destruction in central and southern California because of the low pop-ulation density at the time. A reconstructed isoseismal map for this event is shown in Figure 9.6. A maximum intensity of IX was assigned to the presumed epicentral area near Fort Tejon because of the collapse and destruction of adobe buildings and the overthrowing of trees. In Los Angeles the earthquake shaking frightened people but did not produce any building damage other than super-ficial cracking. An intensity value of VI seems appropriate for Los Angeles. The area shaken at an intensity of VII or greater was ~44 300 km$^2$ suggesting a mag-nitude of about 7.8. The *limit of perceptibility* is only approximately known but the event was felt as far east as Las Vegas. By comparison, the 1906 San Francisco earthquake was reported as felt over a wider area, presumably the direct result of an increase in population density (Figure 9.7).

Of particular note is the record left by surface rupture along a 360 to 400+ km segment of the San Andreas fault in central California. The rupture extended from the San Benito county line southeasterly to the vicinity of San Bernardino as shown by the solid line in Figure 9.7. Fault slip along the rupture zone, however, was non-uniform, but relatively constant along several segments in central California. The slip ranged from 3 to 9.5 m along the central portion of the San Andreas fault (Figure 9.8). Estimates of the seismic moment, which is

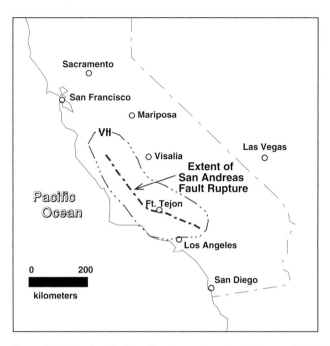

**Figure 9.6** Extent of faulting for the earthquake of January 9, 1857. The dashed contour delineates the approximate area of damaging intensities of VII or more. Data taken from Agnew and Sieh (1978).

proportional to the product of the faulted area and the average fault slip in the individual segments, range from 5.8 to 8.7 × 10$^{20}$ N m depending on whether a depth-of-fault movement of 10 or 15 km is assumed. These values point to a moment magnitude, $M_w$, of 7.8 to 8.0 for the 1857 earthquake. The average fault slip, along a similar rupture length, for the 1906 San Francisco earthquake does not exceed 4 m. A comparable calculation for the 1906 earthquake gives a value for $M_w$ ranging from 7.7 to 7.8 emphasizing that the 1906 event was smaller compared to the 1857 event.

At the Wind Gap Site at Olema, northwest of San Francisco (Figure 9.5), fault trenching observations (Niemi and Hall, 1992) were made across the San Andreas fault segment that ruptured in 1906. A buried Holocene stream channel dated 1800 ± 78 years BP was offset 42.5 ± 3.5 m. This observation indicates a late-Holocene slip rate of 24 ± 3 mm per year for the north coast segment of the San Andreas fault. This slip rate is 40 percent greater than that inferred along the Peninsula segment at Filoli to the southeast. One observation to be emphasized is that since the slip rate is variable along the trace of the San Andreas fault the fault is segmented in its behavior. The implications for estimating the recurrence time of large events are twofold. If the overall slip rate along the San Andreas

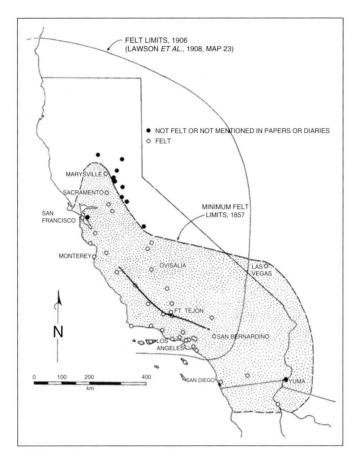

**Figure 9.7** Comparison of the felt areas of the 1857 and 1906 San Andreas fault earthquakes (Sieh, 1978b).

fault is uniform when averaged over long periods of time, then those segments with lower offsets must play catch up in the form of significant earthquakes or be slipping aseismically. Conversely, the long-term slip rate might itself be variable along different segments of the fault zone. This would make generalizations and statements about the average repeat times of a characteristic earthquake more complex.

One conceptual characteristic earthquake model is called the time predictable model. In this model the amount of displacement or fault slip (proportional to the size of the event) in essence specifies the time interval to the next event, provided an estimate of the long-term slip rate is known. Stated in another way, one anticipates that the sum of the individual slips on fault segments measured over many earthquake cycles must balance or be consistent with the plate tectonic slip rate. Given that we have a limited historical record, of at

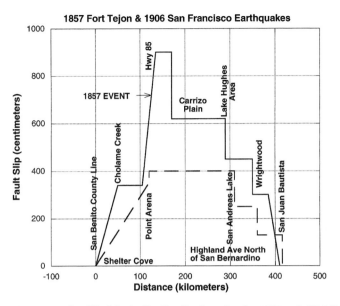

**Figure 9.8** Simplified fault slip distributions for the 1857 and 1906 California earthquakes shown at the same scale. The 1906 earthquake slip is shown by the dashed curve extending from Shelter Cove to San Juan Bautista. Data replotted from Sieh (1978b).

best one or two earthquake cycles, and that plate tectonic slip rates may not be uniform in the short term, the continuing need for data of past earthquake occurrences becomes paramount. It is clear that archaeology can play a role in extending the historical record of earthquake occurrences.

## 9.5    Owens Valley, California earthquake of March 26, 1872

In the very early morning of March 26, 1872, surface faulting and ground rupture took place on the Owens Valley fault along the eastern front of the Sierra Nevada mountain range. The isoseismal map for this event is shown in Figure 9.9. A Mercalli intensity of X was assigned to the vicinity of Lone Pine because every adobe, brick, and stone building in the region was wholly or partially destroyed. Unlike the 1857 and 1906 California earthquakes in which the ground displacements were primarily horizontal, the movement that occurred during the 1872 event was a composite of horizontal and vertical offsets. Horizontal offset averaged about 6 m whereas the vertical movement was approximately 1 m.

Detailed geological mapping showed that the length of the 1872 surface fault rupture was 100 ± 10 km and that the average amount of net oblique slip was 6.1 ± 2.1 m (Beanland and Clark, 1994). Adopting a fault width (the depth along

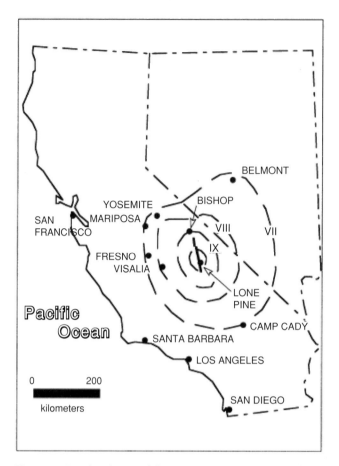

**Figure 9.9** Isoseismal map of the Owens Valley earthquake of 1872. Heavy solid line shows extent of surface faulting. Innermost contour encloses Mercalli intensity of X. Data taken from Oakeshott *et al.* (1972).

the nearly vertical fault plane) of 10 to 12 km yields values of the seismic moment ranging from $1.8 \times 10^{27}$ to $4.4 \times 10^{27}$ dyne cm. These values subsequently give estimates of the moment magnitude of the 1872 earthquake of 7.5 to 7.7, smaller than the 1857 and 1906 events which took place on the San Andreas fault.

It is of some interest to compare this geologically inferred earthquake magnitude from that estimated from the maximum dimension enclosed by the contour of Mercalli intensity VII. Using the linear dimension of 400 km from Yosemite to Camp Cady (an old army post) yields a magnitude estimate of 7.8 (Figure 12.4). Considering the low population density at the time and the subsequent lack of personal accounts, the agreement in the inferred magnitudes is quite good, emphasizing that determinations using historical isoseismal observations are valid.

# Earthquakes of the North American Cordillera

## 10.1 The Basin and Range Province and the Intermountain Seismic Belt

Large earthquakes occur from time to time in the interior regions of western North America. These events take place within two bands of seismic activity, the Eastern California–Central Nevada Zone and the Intermountain Seismic Belt (Figure 10.1). These earthquakes have left many spectacular geological signatures, primarily fault scarps. Fault scarps formed in such a desert-like environment are excellent paleoseismic indicators. These relics of earlier events can easily be recognized hundreds to thousands of years later. However, a steep fault scarp, left by a primarily vertical dip-slip fault that ruptures the surface, would erode and degrade with time. Geologists can estimate the age of scarp formation and hence the time of earthquake occurrence. Such evidence points to earthquake recurrence times of many hundreds to thousands of years for earthquakes in this region.

The December 16, 1954 Fairview Peak–Dixie Valley earthquakes of surface wave magnitudes 7.1 ($M_w = 7.2$) and 6.8 were preceded by two earthquakes of magnitude 6.6 and 6.8 (Rainbow Mountain). These earthquakes also produced surface rupture in a zone about 30 km west of the Fairview Peak–Dixie Valley ruptures. Even though the surface trend of these ruptures was variable, the direction of fault grooves or slickensides left in the bedrock shows a component of both strike- and dip-slip movement; this indicates a consistent west-northwestward direction of extension occurring over a large area of the Basin and Range Province of Nevada. Curiously, even though the 1954 earthquakes produced spectacular vertical fault scarps ranging from 1.8 to 6 m, regional newspapers such as the *San Francisco Examiner* did not dwell on the occurrence of major earthquakes in

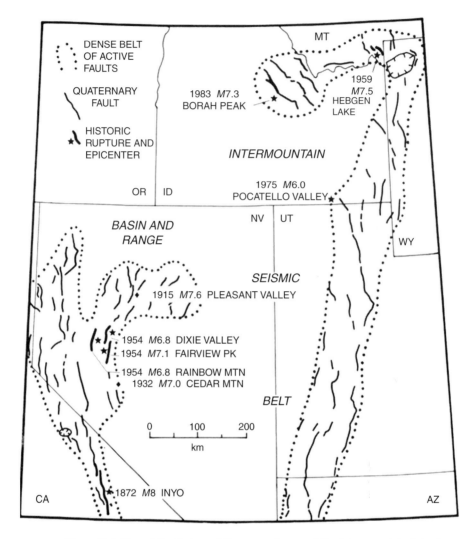

Figure 10.1 Map of the Basin and Range province and the Intermountain Seismic Belt of the western United States showing the locations of the principal seismic events.

Nevada but only described the insignificant ground shaking that was felt in San Francisco.

Fault scarps that formed as a result of the $M_s = 7.6$, October 2, 1915 earthquake in Pleasant Valley, Nevada, are the northernmost scarps created in Nevada in historical time. One of the discontinuous segments of the fault scarp, on the east side of the northerly trending Pleasant Valley, was photographed 8 days after the occurrence of the earthquake (Figure 10.2). Four main scarps were formed in a right-stepping *en echelon* pattern over a combined length of 59 km. The average

Figure 10.2 Fault scarp on the east side of Pleasant Valley, Nevada; photographed on October 10, 1915 (Berry, 1916). Height of scarp about 3 m.

vertical displacement was 2 m and the maximum observed displacement was 5.8 m. Where exposed, the fault plane strikes about N 20° to 25° E and dips westerly between 45° and 80° compatible with regional extension bearing N 65°–70° W. If this earthquake had taken place in a densely populated region there is no doubt that it would have been ranked of far greater destructive historical importance.

A striking regional secondary effect was the large reported increase in the flow of water in streams and springs throughout northern Nevada. In the few cases that were investigated it was found that the increase in flow was the result of secondary cracks that had formed in sand and gravel aquifers (Jones, 1915). It was subsequently reported that after the earthquake over 50 applications for new water-rights were received by the office of the State Engineer of Nevada. Because the 1915 earthquake left especially well-preserved scarps, it is possible to use their dimensions to estimate the seismic moment and moment–magnitude, $M_w$ of the earthquake. Using a mean length of 59 km, an average vertical offset of 2 m, and a down-dip dimension of 17 km, Wallace *et al.*, (1984) determined an $M_w$ of 7.2 for the Pleasant Valley earthquake.

On October 17, 1959, late in the evening local time, a magnitude 7.5 earthquake struck in a rugged mountain region of Montana. This earthquake, known as the Hebgen Lake event, is the largest earthquake recorded in the

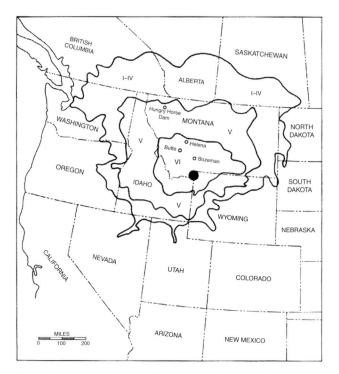

**Figure 10.3** Isoseismal map of the $M = 7.5$ Hebgen Lake, Montana earthquake of August 17, 1959. Intensities within filled circle ranged from VII to X.

Intermountain Seismic Belt. The earthquake was perceptible over an area of 1 500 000 km² (Figure 10.3). A maximum intensity of X was very local however, confined to the meizoseismal area where surface fault rupture occurred (Figure 10.4). The Mercalli intensities fell off very rapidly from this zone and would not have exceeded a value of VIII if estimations were based on vibration effects alone. At Bozeman, Montana, 93 km from the epicenter, the maximum recorded ground acceleration was only 7 percent of gravity. The area subjected to an intensity of VII or greater was limited to an elliptical area of about 6200 km². Much of the damage was attributed to a large rockslide and secondary landslides (Figure 10.5). The Hebgen Lake event differs in one respect from the Nevada earthquakes of the Basin and Range Province in that motion on the fault rupture plane was pure dip-slip normal faulting.

First-hand accounts of fault scarps being formed and observed during an earthquake occurrence are infrequent. A notable exception took place during the magnitude 7.3, Borah Peak, Idaho earthquake on October 28, 1983. Two hunters were traveling along a dirt road when the earthquake struck producing violent shaking for a duration of 10 to 15 s. The observers were on the downthrown side of a fault scarp formed 20 m in front of their vehicle. The vertical offset on the

**Figure 10.4** View of the Hebgen Lake, Montana, fault scarp. Vertical scarp on hillside northeast of Hebgen Lake trends southeastward. Height of the scarp is about 3 m for most of its 10 km length (photograph by Bob Wallace, US Geological Survey).

**Figure 10.5** Landslide on the northeast shore of Hebgen Lake, which cuts across Montana State Route 287 (photograph by Bob Wallace, US Geological Survey).

**Figure 10.6** Fault scarp formed 20 m in front of two hunters traveling on a dirt road as a result of the Borah Peak earthquake of 1983. The height of the scarp is 1.8 m. White dashed line represents forward edge of rupture zone (Pelton *et al.*, 1984).

scarp was 1.8 m (Figure 10.6). Of particular interest is the fact that the observers described the formation of the scarp as taking place over a time span of 1 s (Pelton *et al.*, 1984; Wallace, 1984). The moment magnitude of this earthquake was subsequently determined to be 7.0.

## 10.2     Sonora, Mexico earthquake of May 3, 1887

The American southwest has from time to time been subjected to earthquakes that have produced seismic intensities of VII or greater (Table 10.1). One of the largest earthquakes to have affected the southwestern United States and northern Mexico took place on May 3, 1887. Because the area was sparsely populated, the number of fatalities was small but property damage was extensive. The isoseismal map (Figure 10.7) shows maximum intensities of X to XI in the meizoseismal area. The area experiencing intensities of VII or greater had dimensions of 575 km by 425 km. Intensities of III were reached as far south as Mexico City. The presumed epicenter was south of the border between the state of Arizona and Mexico, near the small town of Colonia Morelos (Figure 10.8). Segmented surface fault rupture (Suter and Contreras, 2002) extended about 100 km southward from a point beginning about 8 km south of the international border. At Bavispe the passage of the seismic disturbance was described as coming from the northwest, implying a rupture front passing to the southeast. The asymmetric shape of the isoseismals suggests a directivity or focusing of seismic intensity to the southeast. The average vertical displacement on the fault rupture was 2

Table 10.1. *Some significant earthquakes of the American southwest (excluding California).*

| Date | Location | Comments |
|---|---|---|
| 1830 | 31.9° N, 110.1° W | Precise date unknown. Intensity VII to IX in San Pedro Valley, Arizona. |
| May 3, 1887 | 30.8° N, 109.25° W | Described in Section 10.2. |
| January 25, 1906 | 35.2° N, 111.7° W | Intensity VII at Flagstaff, Arizona. |
| July 12, 1906 | 34.1° N, 106.9° W | Intensity VII at Socorro, New Mexico. |
| July 16, 1906 | 34.1° N, 106.9° W | Intensity VIII at Socorro, New Mexico. $M = 6.5$. |
| November 15, 1906 | 34.1° N, 106.9° W | Intensity VIII at Socorro, New Mexico. |
| April 7, 1908 | 30.63° N, 109.47° W | Intensity VI at Fronteras, Sonora. Estimated $M = 4.8$. |
| May 26, 1907 | 30.8° N, 109.3° W | Intensity VIII at Colonia Morelos, Sonora, Mexico. $M = 5.2$. |
| September 24, 1910 | 35.75° N, 111.50° W | Intensity VII at Cedar Wash, Arizona. |
| August 18, 1912 | 35.95° N, 111.95° W | Intensity VII to VIII at Lockett Tanks, Arizona. $M = 5.5$. |
| May 28, 1918 | 35.4° N, 106.2° W | Intensity VIII+ at Cerillos, New Mexico. |
| December 18, 1923 | 29.9° N, 109.3° W | Intensity IX at Huasabas, Sonora, Mexico (~100 km SW of Bavispe). $M = 5.7$. |
| August 16, 1931 | 30.6° N, 104.5° W | Intensity VIII at Valentine, Texas. |
| September 17, 1938 | 33.2° N, 108.6° W | Intensity VII at Silver City, New Mexico. |
| September 20, 1938 | 33.25° N, 108.6° W | Intensity VII at Buckhorn, New Mexico. |
| May 4, 1939 | 36° N, 114.8° W | Intensity VII at Hoover Dam, Arizona. $M = 5.0$. |
| January 17, 1950 | 35.71° N, 109.54° W | Intensity VII at Ganado Trading Post, Arizona. |
| July 21, 1959 | 36.8° N, 112.37° W | Intensity VI to VII at Fredonia, Arizona. $M = 5.5$ to $5.75$. |
| September 11, 1963 | 33.2° N, 110.7° W | Intensity VI at Globe, Arizona. $M = 4.1$. |
| October 28, 1965 | 29.4° N, 107.9° W | $M = 5.0$. |
| January 22, 1966 | 37.0° N, 107.0° W | Intensity VIII at Dulce, New Mexico. $M = 5.5$. |
| December 25, 1969 | 33.4° N, 110.6° W | Intensity VI at San Carlos, Arizona. $M = 4.4$. |
| December 4, 1971 | 35.2° N, 112.2° W | Intensity V at Williams, Arizona. $M = 3.7$ (Brumbaugh, 1980). |
| February 4, 1976 | 34.66° N, 112.5° W | Intensity VI at Chino Valley, Arizona. $M = 5.1$ (Eberhart-Phillips *et al.*, 1981). |

to 4 m with the western side downthrown (Figure 10.9). Gianella (1960) noted additional right-lateral offset of up to 6 m cutting through volcanic rocks near Bavispe. This is compatible with a regional northwest–southeast orientation of tectonic extension. Such a direction of extension would produce roughly comparable components of strike-slip and dip-slip offset on a northerly trending fault. A moment magnitude of 7.4 has been estimated from the faulting dimensions

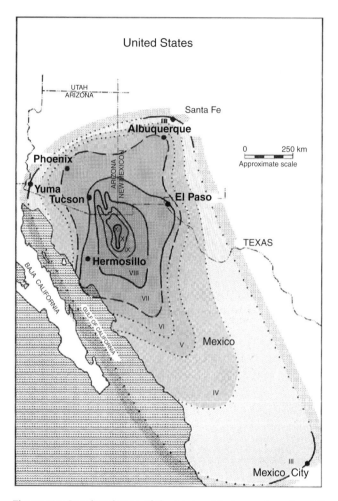

**Figure 10.7** Isoseismal map of the Sonora, Mexico earthquake of May 3, 1887. Maximum intensities of X to XI were achieved (from DuBois and Smith, 1980).

(Wells and Coppersmith, 1994), making the Sonora event one of the largest known for the American southwest, excluding California. Adobe structures and the church collapsed in Bavispe. Failure of structures was the direct result of wall separation and subsequent roof collapse. Roofs were heavy and constructed by laying log rafters called *vigas,* which were untied from wall to wall; the rafters were covered with brush or cane and mud (Figure 10.10). In Bavispe, walls that were oriented north–south fell to the east, whereas walls positioned east–west fell mainly to the south, roughly consistent with the direction of motion for right-lateral strike-slip on a north-northwesterly trending fault.

The Sonora earthquake is also noteworthy in that it was recorded and noted by several Indian tribes of the American southwest. It needs to be emphasized that

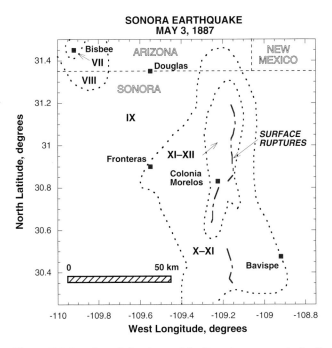

**Figure 10.8** Location of the observed fault rupture segments for the Sonora earthquake of 1887. The total length between the north and south extremities of the rupture is about 100 km. Isoseismals of the epicentral area are shown by the dotted contours. Data from DuBois and Smith (1980) and Suter and Contreras (2002).

**Figure 10.9** Photograph of the 1887 scarp viewed from the west in 1974. The scarp cuts through alluvium and reaches a maximum height of 4 m (Sumner, 1977).

**Figure 10.10** Damage to adobe structures in Bavispe, Mexico, from the 1887 Sonora earthquake. Wooden roof rafters were not tied to adobe walls (AHS 22912, courtesy of the Arizona Historical Society, Tucson).

this earthquake was a significant event inasmuch as the relative importance and documentation of particular events was often viewed differently by the American Indians. Mention was made in the Introduction of the recording of this event on a wooden calendar stick of the Maricopa Indians residing near the junction of the Gila and Salt rivers in Arizona. This event was also found on calendar sticks of the Pima tribe. The Papago Indians are members of the Pima tribe who live in Arizona, west and southwest of Tucson about 100 km into the neighboring state of Sonora, Mexico. A calendar stick examined by Lumholtz (1912, p. 75) was explained by its owner to note that the 1887 earthquake took place during the "flowers disappear" (plants begin to make fruit) moon. The Papago calendar was divided into 13 moons.

A northern group of the Pima Indians occupies lands south of Phoenix, Arizona that straddle the Gila River. Their lands are known today as the Gila River Indian Reservation. A few notched sticks were found among the northern Pimas covering a period of 70 years beginning from the moon that preceded the well-documented meteor shower of November 13, 1833. The Pima year began with the saguaro harvest in about June and years were marked by a deep notch carved across the stick. An Indian named Owl Ear stated that "it [earthquake of May 3, 1887] was noticed by many of our people, if not by all, who wondered why the earth shook so" (Russell, 1908, p. 60).

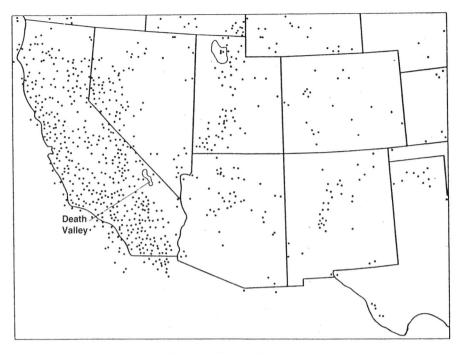

**Figure 10.11** Location of Death Valley in relation to seismic epicenters in the southwest United States (taken from Woollard, 1958).

## 10.3    Death Valley, California earthquakes

Death Valley, which lies in southeastern California, is one of the valleys of the western Basin and Range Province. It lies about halfway between the Sierra Nevada and the Colorado River and north of the Mojave Desert. The mountains bordering the main valley are northerly trending fault blocks. Faults in the Death Valley region exhibit many classic geological features such as great length and linearity, scissoring of Quaternary fault scarps, and horizontally offset rock units and stream channels. Despite the presence of youthful geomorphic features and the fact that many of the mapped faults are therefore geologically young, the level of current seismic activity is surprisingly low. One could easily describe Death Valley as an island of tranquillity in a sea of seismicity (Figure 10.11).

Epicenters of earthquakes located in the Death Valley region that occurred from 1939 to 1999, are shown in Figure 10.12. In compiling the epicenter map the locations used were those reported by the network operated by the California Institute of Technology from 1932 onward (later known as the Southern California network, operated in conjunction with the US Geological Survey) and the network operated by the University of California at Berkeley since 1910. In

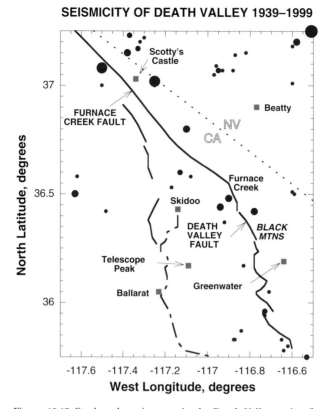

**Figure 10.12** Earthquake epicenters in the Death Valley region from 1939 to 1999.
Size of circles is proportional to magnitudes binned at 0.5 magnitude intervals.
Largest circles are for magnitudes of 5.1 to 5.5, smallest circles represent
magnitudes of 3.1 to 3.5.

addition, use was made of microearthquake locations reported from the southern Great Basin network operated by the US Geological Survey from 1981 to September 1992. The spatial station coverage has not been uniform with time but all reported shocks of magnitude 3.0 or greater are shown. Several observations can be made from the seismicity map. Much of the Furnace Creek fault zone appears to be relatively quiet compared to the Death Valley fault zone. Larger-magnitude shocks appear to cluster in the vicinity of Scotty's Castle, near the confluence of the Furnace Creek and Death Valley faults.

On the other hand, the paleoseismic manifestation along much of the Furnace Creek–Death Valley fault system is spectacular. The last major earthquake in Death Valley took place 2000 years ago with displacement occurring along the Death Valley fault at the western edge of the Black Mountains (Hunt, 1975). Prominent vertical escarpments of 3 to 6 m can be seen along the length of the fault from near Furnace Creek southward for a length, $L$, of about 43 km. Assuming a depth of faulting of 15 km, the estimated seismic moment of this event was $5.8 \times 10^{19}$ N m. This corresponds to a moment magnitude, $M_w$, of 7.2 and compares well with estimates based on empirical correlations of fault length, $M = 6.1 + 0.7 \log L$.

The largest earthquake in the Death Valley region in the twentieth century took place in the early hours of November 4, 1908. Richter (1958a) assigned a *questionable* epicentral location of 36° N, 117° W, and a magnitude of 6.5. It is interesting to re-examine the available data. The following description from Townley and Allen (1939) is informative:

> 1908 November 4, 12:37 a.m. VIII (Rossi–Forel). Death Valley, Inyo County. Press dispatches from San Bernardino on November 10 and 11 contained accounts of shocks which had disturbed the Death Valley region for about three weeks, all of them slight except one great shock in the early morning of November 4. The account, appearing to be exaggerated but founded upon fact, said the continuance of the shocks, together with the violence of that of November 4, had caused many prospectors to leave the region, fearing the return of conditions when the valley had been under water. Since there were no cities and very few buildings within many miles of the point of origin of this shock, it is impossible to more than guess at its intensity, but it was sufficiently strong to record on the seismographs of the day to considerable distances, the recording at Ottawa continuing for twenty-six minutes [probably surface waves on an undamped seismograph].
>
> The shock commenced to record on the Berkeley seismograph at 12:37:39 a.m. [Records cannot now be located at Berkeley].
>
> At Lone Pine the observer of the Weather Bureau reported two heavy shocks at 12:46 a.m., one sharp, and the other rolling. He dated the entry November 3, evidently an error. At Independence E. W. Brooks, Weather Bureau observer, reported three shocks during the night, the first at 11:50 p.m. November 3, the second at 12:20 a.m., November 4, the third without giving the time [some discrepancies with arrival time at Berkeley]. Intensity was VI, motion north–south, and duration six seconds.
>
> The shock was felt at Tehachapi, Kern Co., and probably in San Bernardino.

In assessing the reported seismic intensities for this shock it is important to be aware of the population density in the Death Valley area at the time. The West's last big gold rush between 1905 and 1908 brought several thousand people into the Death Valley region (Lingenfelder, 1986; Nadeau, 1965). The short-lived mining towns of Greenwater and Skidoo, each with their own newspapers, the *Greenwater Times* and the *Skidoo News*, appeared in the mountains above Death Valley. Both newspapers folded in the summer of 1908 along with the mining boom and unfortunately could not be used for any local reports of the November, 1908 shock. No felt reports were found in the October to November 1908 copies of the *Rhyolite Herald*, the *Rhyolite Daily Bulletin*, and the *Las Vegas Age* (Lingenfelder and Gash, 1984); copies of these papers were held in the archives of the Nevada Historical Society.

There were also no reports in the *Inyo Register (Bishop)* of the November 4 earthquake being felt at Bishop, California. However, the issue of December 3 reports on Death Valley earthquake activity: "Eugene E. Smith says that the stories of repeated earthquakes in the Death Valley country are not all fakes by any means. He has lately been out that way, and in places great masses of rocks have been thrown into the canyons, in some cases seriously interfering with the roadway." The following description of the earthquake by a miner who reached San Bernardino was reported on November 13–14 in the *Goldfield Chronicle* and the *Goldfield Daily Tribune*:

> The dismal crags of the Funeral Range seemed to fairly totter when the severe shock came last week. Miners were tossed from their bunks, camp equipment scattered about, and immense boulders thrown down. Horses and mules stampeded, and they were quickly followed by the men, who gathered up their portable effects and left for more restful lands, leaving behind scores of claims and rich prospects.

The description is supportive of a Modified Mercalli intensity of VI to VII in the meizoseismal region. No damage was reported in Lone Pine and Independence. The lack of damage at Lone Pine suggests that the Modified Mercalli intensity value cannot have exceeded V. Bearing in mind the questionable epicentral location and the absence of felt reports at the populated towns of Bishop, California, and Rhyolite and Las Vegas, Nevada, the isoseismal map of Figure 10.13 should be viewed as speculative. The area enclosed by the intensity contour of V is estimated not to exceed 50 000 km$^2$. The empirical relationship between seismic intensity and earthquake magnitude (Toppozada, 1975), calibrated for California and western Nevada earthquakes, yields a magnitude of ~6 for the November, 1908 Death Valley earthquake.

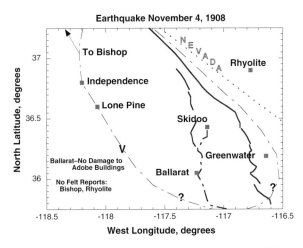

**Figure 10.13** Estimated isoseismal map for the Death Valley earthquake of November 4, 1908.

Geological, geodetic, and seismic data show that Death Valley is being widened in a northwesterly–southeasterly direction. Movement is taking place along the northwest-trending strike-slip Furnace Creek fault zone and by regional extension taking place by oblique slip on north-trending mountain range frontal faults. The paucity of cumulative seismic activity in the region, however, needs to be addressed. Global positioning observations show that right-lateral movement is taking place along the Furnace Creek–Death Valley fault zones at a rate of about 3 mm per year (Gan *et al.*, 2000). Normal slip is undoubtedly also taking place along this fault system but, except for inferences from the Black Mountain event 2000 years ago, data on vertical deformation rates are lacking.

For a 100-year interval, a slip rate of 3 mm per year corresponds to a cumulative slip of 30 cm. With an assumed total fault area of 200 km $\times$ 15 km, the equivalent seismic moment is $2.7 \times 10^{19}$ N m or one $M_\mathrm{w} = 6.9$ earthquake. An event of this size in the past 100 years cannot be found in the historical record for Death Valley. A significant amount of deformation must be taking place *aseismically* or it may be that large seismic events cluster in time at unknown intervals.

# Earthquakes of eastern and central North America

## 11.1 Great Quebec earthquake of 1663

The 1663 La Malbaie, Quebec, earthquake was one of the first great events in North America for which a written record is preserved. The descriptions of its effects were undoubtedly exaggerated but shed some insight into early American Indian beliefs. "The Indians [Algonquians of the Ottawa River Valley] discharged their firearms in order to drive away the *manitou*." (According to Algonquian conception manitou was a supernatural force that pervaded the natural world.) The La Malbaie, Quebec, earthquake had an inferred magnitude of ~7 with several aftershocks and occurred late in the afternoon of February 5, 1663. As chronicled in the Jesuit Relations of 1663 (Thwaites, 1896–1901): "It began half an hour after the close of Benediction on Monday, the 5th of February, the feast of our holy martyrs of Japan, namely at about 5-1/2 o'clock, and lasted about the length of 2 *misereres*." (A *miserere*, the 51st psalm of the Bible, takes about 1 min 30 s to recite. It begins "Have mercy upon me, O God . . . ," perhaps an appropriate pleading during an earthquake.) Remembering the sparseness of the population and that transportation and intercommunications were limited, descriptions were embellished and amplified with the passage of time. The descriptions of mountains slipping, tall trees clashing, fence posts jumping, etc. are probably the statements of frightened observers who had not as yet experienced an earthquake. Nevertheless, the earthquake was widely felt over much of northeastern America, from Montreal to Boston. It is not possible to be precise about its exact epicentral location but it is assumed to lie below Quebec City in the St. Lawrence River at the same location as the earthquake of February 28, 1925. The distribution of seismic intensities for the 1663 shock is believed to

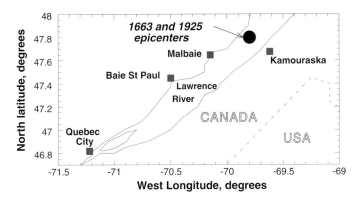

Figure 11.1 Map showing the epicentral areas of the 1663 Quebec and 1925 St. Lawrence earthquakes.

Figure 11.2 Row of cemetery monuments in Riviere Ouelle cemetery, Canada, all fallen to the southeast as a result of the 1925 St. Lawrence earthquake (Hodgson, 1925b).

be similar to that of the later 1925 shock, which was studied in some detail by Hodgson (1925a, b, 1928, 1950). Both of these earthquakes were centered about 120 km northeast of Quebec City in a currently active seismic zone of 150 km × 100 km centered near Malbaie (Figure 11.1).

Shaking from the 1925 St. Lawrence, or Charlevoix–Kamouraska earthquake, was felt at distances more than 1000 km from the epicenter. Damage descriptions indicate that an elliptical area of about 20 000 km$^2$, oriented with its maximum axis along the trend of the St. Lawrence River was subjected to a Modified Mercalli intensity of VIII. In Quebec City and Shawinigan Falls, hundreds of kilometers to the southwest, some building damage was reported. The damaged structures had been built on deep soft soil and clay deposits, which are known to amplify earthquake ground motions.

One of the interesting observational facets about the 1925 St. Lawrence earthquake was the large number of overturned and rotated cemetery monuments in the epicentral area (Figure 11.2). Subsequent seismological studies of the 1925 earthquake reveal that its moment magnitude was 6.8. The radiation pattern of seismic waves suggests that the mechanism of failure was thrust (reverse) faulting on northeast–southwest nodal planes that dip 45°. Data are consistent with a maximum compressive principal stress that is horizontal and is oriented in a northeast-southwest direction in this part of Canada. Earthquake activity currently occurs along the axis of the St. Lawrence River and failure along the existing zone of weakness is probably controlled by intraplate convergence forces.

Table 11.1 lists other earthquakes with a magnitude of 5 or greater that have taken place in northeastern America during the seventeenth century. Most of these events cannot be classified as great events but they do emphasize that moderate earthquakes occur from time to time in this region.

## 11.2    Charleston, South Carolina earthquake of 1886

On August 31, 1886, the largest earthquake ever reported in the eastern region of the United States struck in the vicinity of Charleston, South Carolina. The earthquake was felt as far away as Boston, the upper Mississippi region, the islands of Cuba, and Bermuda. The felt reports indicate a radius or *limit of perceptibility*, taken as intensity II on the Mercalli Intensity Scale, of approximately 1500 km (Figure 11.3). This earthquake occurred before instrumental recordings and therefore its moment magnitude was estimated to be 7.3 using empirical relationships between felt areas, various Modified Mercalli intensity isoseismal areas, and moment magnitudes for earthquakes in North America (Johnston, 1966b). This value, however, may seem low considering the size of the felt area. Because of the different geological characteristics and attenuation properties of the eastern United States, the area of damage for earthquakes of the same magnitude is larger for eastern and central United States earthquakes than for western United States earthquakes.

Table 11.1. *Seventeenth-century eastern North American earthquakes*[a]

| Date | Location | Maximum Modified Mercalli intensity (MM) or estimated magnitude |
|---|---|---|
| June 11, 1638 | St. Lawrence Valley. Felt from Quebec to New England | MM VIII to IX, $M = 6.5 \pm 0.5$ |
| June 12–July 1, 1638 | New England | MM $\leq$ III |
| January 24, 1639 | New Hampshire | MM III |
| March 15, 1643 | Northeastern Massachusetts. Felt at Salem, Boston, Danvers, Lynn, and Haverhill | MM III to V |
| December 1660 | Roxbury, Massachusetts | |
| February 11, 1661 | Trois-Rivieres, Canada. Felt at Boston | $M > 5$ |
| February 5, 1663 | La Malbaie, Canada | MM IX to X $M = 7.0 \pm 0.5$ |
| February 6–7, 1663 | La Malbaie, Canada | 4 events of $M > 5$ |
| 1664 | St. Lawrence Valley, Canada | Probable aftershock |
| February 24, 1665 | Tadoussac, Quebec | MM VIII? |
| October 16, 1665 | Quebec City, Canada | $M > 5$ |
| January 9, 1667 | Roxbury, Massachusetts | |
| April 13, 1668 | Damage at Cape Tourmente, Canada | $M > 5$ |
| December 19, 1668 | Northwest of Boston, Massachusetts | $M = 3.5$ |
| November 30, 1669 | Roxbury, Massachusetts | |
| February–March 1672 | Tadoussac, Canada | Aftershocks of 1663? |
| December 8, 1673 | Tadoussac, Canada | |
| February 13, 1678 | Stamford, Connecticut | |
| June 30, 1678 | Stamford, Connecticut | |
| September 12, 1682 | Felt at Boston, Massachusetts | |
| February 18, 1685 | Cape Ann, Massachusetts | $M = 3.5$ |
| September 8, 1688 | Bristol, Connecticut | |
| November 15, 1690 | Dover, New Hampshire | |
| October 25, 1693 | Dover, New Hampshire? | |
| January 31, 1694 | Dover, New Hampshire | Aftershock? |
| February 20, 1697 | Brockton, Massachusetts | $M = 3.0$ |
| January 20, 1699 | Danbury, Connecticut | |
| February 10, 1700 | Dover, New Hampshire | |
| February 11, 1700 | Dover, New Hampshire | Doubtful event |
| March 11, 1700 | Newbury, Massachusetts | |

[a] Stevens (1995) and Ebel (1996) list original sources. Many other reported events were found to be fictitious or incorrectly dated according to the Julian calendar rather than the Gregorian calendar. In North America reported dates prior to 1752 have to be carefully scrutinized. Even though the change from a Julian calendar to a Gregorian calendar was ordered by Pope Gregory in 1582, to correctly allow for the true length of the year, most Protestant nations and the Greek Church did not recognize papal mandates. The change was not implemented by England and her colonies until 1752.

**Figure 11.3** Isoseismal map of the Eastern United States contoured to show the broad regional patterns of the reported intensities for the 1886 Charleston earthquake. Contoured intensity levels of the Modified Mercalli Scale are shown in roman numerals (taken from Bollinger, 1977).

The Charleston earthquake was an unusual occurrence in several aspects. Firstly, the earthquake took place in a region previously free of notable seismic events since the English arrived in 1670. Earthquakes of great size would not have escaped the notice of the early east-coast centers of population. Thousands of aftershocks, including earth noises, lasted for eight years or more. The three largest aftershocks included two large earthquakes on October 22, 1886, and one on November 5, 1886. Magnitudes of these aftershocks were inferred to be 5.1, 5.7, and 5.3 (Talwani and Sharma, 1999). Within the epicentral area much of the ground is underlain by sands that are saturated with water. As a result portions of the meizoseismal area were subjected to liquefaction and the formation of water, sand, and mud fountains or blows, and subsequently covered with sand. Spectacular sand craterlets were observed (Figure 11.4). Paleoliquefaction investigations in the coastal plain of South Carolina suggest a recurrence time of 500–600 years for $M = 7+$ earthquakes in the vicinity of Charleston (Talwani and Schaeffer, 2001). It is notable that the seismic intensity reached a value of

**Figure 11.4** A typical sand crater produced by the throwing out of water and sand, from a few meters below the surface, as a result of the Charleston earthquake (Dutton, 1889).

X to XI in the city of Charleston but there were only 60 fatalities. Many brick buildings still standing today withstood the effects of the earthquake. Much of the lack of damage to unreinforced brick buildings, which typically do not fare well during earthquake shaking, was attributed to the careful workmanship in the bonding and the quality of the mortar used. The bricks were hand-made with somewhat rough surfaces and the mortar, made of burned lime from well-sorted oyster shells (tabby), adhered with great tenacity.

The meizoseismal area of the Charleston earthquake is close to the Atlantic coastline but well within the North American plate (Figure 11.5). Current thinking (Talwani, 1999) is that the earthquake(s) in 1886 were associated with composite motion along the northwest-trending Ashley River fault zone and the north-northeast-trending Woodstock fault zone. Both of these zones are delineated by the pattern of current seismic activity and are being subjected to maximum horizontal compressive stresses oriented in a northeast–southwest direction. Fort Dorchester, a structure built on the banks of the Ashley River in 1775, and located in the epicentral area, shows left-lateral fault offset crossing the walls of the fort (Figure 11.6).

## 11.3    New Madrid earthquakes of 1811–1812

Three earthquakes with magnitudes near or exceeding 8, and thousands of aftershocks lasting more than five years, struck the central Mississippi Valley

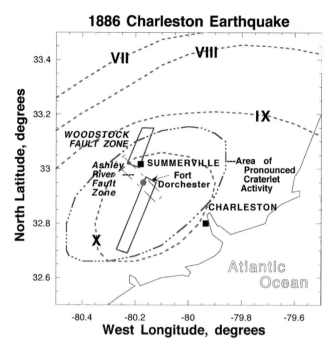

**Figure 11.5** Isoseismals of the 1886 Charleston earthquake. Area of pronounced craterlet activity produced by liquefaction is also shown. The pattern of current seismic activity is associated with the north-northeast-trending Woodstock fault zone and the north-west-trending Ashley River fault zone. The Ashley River fault intersects and offsets the Woodstock fault zone in a left-lateral motion.

near New Madrid, Missouri in the winter of 1811–1812. The sequence began on December 16 in the early hours of the morning and was notable for the large area of damage, even though there were few settlements near. Felt accounts were reported on the Gulf Coast, the Atlantic shore, and in Quebec to the northeast. There are no documented felt reports from the west. The large damage area (an estimated area of 600 000 km² for Mercalli intensity VII) and the large felt area of approximately 5 000 000 km² are striking (Figure 11.7). Because of the low attenuation of the propagation of seismic wave energy, earthquakes are *felt* at greater distances in the eastern and central regions of the United States when compared to the west. No earthquake in the recorded history of North America has had a damage or felt area anywhere near the size produced by the New Madrid earthquakes.

Accounts of the earthquake, experienced by an observer in New Madrid, stressed the violence of the shocks, the disruption of the land surface, and the formation of sand blows:

**Figure 11.6** Two views of the fault offset of the Ashley fault through the northern wall of Fort Dorchester; 10 cm of left-lateral offset can be seen through tabby wall (roasted oyster shells as mortar) 0.76 m thick and 2.1 m high. The crack is also visible in the southern wall indicating offset along a bearing of N 20° W.

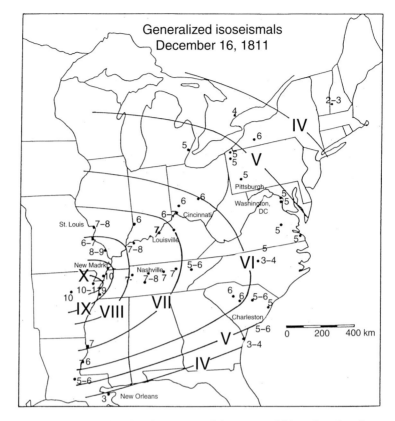

**Figure 11.7** Generalized isoseismal map of the New Madrid earthquake of December 16, 1811. The isoseismals indicate the outer bound of the region of specified Modified Mercalli intensity (Nuttli, 1973).

On 16th Dec., 1811, about 2 A.M. we were visited by an earthquake shock accompanied with an awful noise . . . there were several lighter shocks . . . until the 23rd of January 1812, when a shock was felt that was as violent as the most severe of the former shocks . . . on the 7th [of February] at 4 P.M. a concussion took place which was so much more violent than the preceding that it was known as the hard shock . . . the earth seemed horribly torn to pieces. The surface of hundreds of acres was from time to time covered over for various depths by sand which issued from the numerous fissures. Some of these fissures closed up immediately after they had vomited forth sand and water. (Broadhead, 1902, p. 77)

The eruption of sand and water, as a result of the liquefaction of subsurface sand layers, flooded approximately 3300 km² (Figure 11.8). Sand ejected in this manner

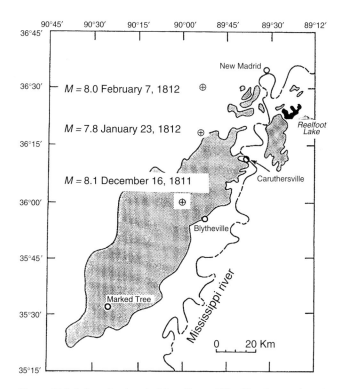

**Figure 11.8** Inferred epicentral locations of the $M = 8+$ earthquakes of the New Madrid sequence and the distribution of subsequent sand blows.

is often deposited around the fissure or spout leaving a permanent miniature crater. Ground uplift and subsidence of several meters took place in the meizoseismal area forming a region known locally as the "Sunken Country." There was subsidence of 3 m and subsequent formation in northwestern Tennessee of Reelfoot Lake with submerged trees. Reelfoot Lake is today about 12 km in length and 3 to 5 km in width. Along the bluffs of the Mississippi River landslides were common.

The New Madrid earthquakes have been discussed in Indian legends and early American folklore. One romantic earthquake legend of note pertains to Reelfoot Lake, on the Tennessee side of the Mississippi River, which was formed as a result of the New Madrid earthquake of December 16, 1811, probably the strongest earthquake ever felt in North America. Reelfoot Lake was created when a swampy tract sank several feet and subsequently filled with water by backflowing of the Mississippi River caused by temporary blockage of the normal drainage channels. The legend is as follows. A handsome chief named Reelfoot was born with a club foot. Reelfoot loved a beautiful Choctaw princess but her father disapproved of the courtship. Reelfoot and his friends abducted the princess and proceeded with the wedding but incurred the wrath of the Great Spirit. During the wedding festivities the Great Spirit stamped his foot causing the earth to shake. As a result the Mississippi River (the "Father of Waters") reversed its course submerging Reelfoot, his new bride, and the wedding party, and forming a new lake.

An Indian prophecy has also been attached to the New Madrid earthquakes. Tecumseh (Tecumtha) was one of the truly great American Indian leaders who preached unification from 1807 to 1812, to resist the enroachment of the white settlers. His famous "prophecy" given presumably at the ancient Creek Indian town of Tukabachi (near the present site of Montgomery, Alabama) is of historical interest:

> You do not believe the Great Spirit has sent me. You shall know. I leave Tuckhabatchee directly, and shall go straight to Detroit. When I arrive there, I will stamp on the ground with my foot and shake down every house in Tuckhabatchee . . . The morning they had fixed upon as the day of his arrival at last came. A mighty rumbling was heard – the earth began to shake; when at last, sure enough, every house in Tuckhabatchee was shaken down. The exclamation was in every mouth, "Tecumthé has got to Detroit!" It was the famous earthquake of New Madrid on the Mississippi. (McKenney and Hall, 1838–1844; Mooney, 1896)

A later source describes the same story but states that the earthquake would occur when Tecumseh returned to Tippecanoe, near Lafayette, Indiana (Tucker, 1956). However, history shows that Tecumseh was in neither place when the earthquakes began. Nevertheless, the Creek Indian legend has persisted.

The New Madrid earthquakes also contributed much to the folklore and religious enthusiasm and conversions of the unchurched white settlers of the time. One tale states that through an oversight the Lord forgot about Reelfoot lake during the original creation: "When he finally did remember it, after goodness knows how many thousands of years, he was so put out about it he didn't think about it bein' Sunday, and he just ripped up the earth and made that lake as quick as he could" (Viitanen, 1973).

Some clues as to the underlying cause of the New Madrid earthquakes can be found by examining current seismic activity. Some of this seismic activity is clustered in a north-northwesterly band running southeasterly into northwestern Tennessee. The pattern of crustal stresses, as deduced from earthquake focal mechanisms, shows that the New Madrid seismic zone is being compressed. The compression is oriented in an east-northeast to west-southwest direction and is consistent with internal pressures formed in the North American plate as it moves away from the accreting axis of the Mid-Atlantic Ridge boundary. However,

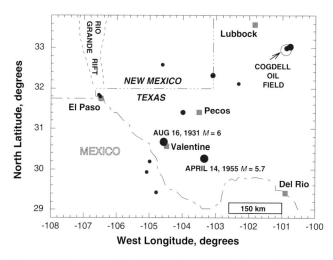

**Figure 11.9** Recent earthquake locations in West Texas. The smallest filled circles are magnitudes 4.0–4.5. Small earthquakes induced by oil-field operations have been associated with the Cogdell oil field.

this simple conceptual model does not explain why earthquakes would be preferentially concentrated in the New Madrid region. Grollimund and Zoback (2001) have argued that melting and removal of the ice sheet, that covered large parts of the northern United States 20 000 years ago, perturbed the regional stress field in the region of the New Madrid epicenter. Their modeling demonstrated that glacial unloading increased the level of differential stress in the crust, in an east–west direction, facilitating the onset of brittle failure. There is no doubt that a repeat in today's environment of an earthquake comparable in size to the 1811–1812 events would produce a major catastrophe. How frequently should we expect events of this size to recur in the central United States? An indication can be obtained from frequency–magnitude considerations. Nuttli (1974) examined an earthquake catalog for the central Mississippi region covering the years 1833–1972. Even though the data set is probably not complete (many early events may have escaped detection) straightforward extrapolation suggests that earthquakes of New Madrid's size have a recurrence time in the range of 500 to 800 years. Later studies (Newman *et al.*, 1999) that combine global positioning system measurement data with recent seismic data predict recurrence times of magnitude 8 events in the New Madrid zone well in excess of 2500 years, much larger than the previous estimate.

Examination of earthquake-induced liquefaction features, particularly sand blows, associated with archaeological indicators – such as artifacts and botanical and charred material of native American occupation horizons – shows evidence for at least one significant earthquake about AD 1300 ± 100 and several earlier events between 3340 BC and AD 780 (Tuttle *et al.*, 1996). Trenching excavations in the vicinity of Reelfoot Lake have revealed liquefaction features supporting a major seismic event in the time frame from AD 1200 to 1650 (Kelson *et al.*, 1996). The geological data therefore suggest a repeat time of 500 to 1000 years for large earthquakes in the New Madrid seismic zone.

Were the 1811–1812 New Madrid events smaller in magnitude than usually assumed? A recent study by Hough *et al.* (2000) examines this question. Arguments are presented, based on a better understanding of local site effects, that the early descriptions of dramatic shaking are not consistent with the felt reports leading to lower estimates of Modified Mercalli intensity values. As a result isoseismal area–moment magnitude estimates of the size of the three principal mainshocks on December 16, January 23, and February 7 give $M_w$ values of 7.2, 7.0, and 7.4, respectively. These values give a combined magnitude estimate for the New Madrid sequence of approximately 7.5.

### 11.4    Texas earthquakes

Earthquakes are felt in Texas from time to time, but the number of damage-producing earthquakes with epicenters in Texas is small. From 1847 to 1986, 24 events have been associated with some reported damage. West Texas has been the location of several earthquakes, most of which have been felt in El Paso. These earthquakes are primarily associated with the extensional stress regime of the Rio Grande rift system, a linear structural depression that extends from southern New Mexico into west Texas and Mexico (Figure 11.9).

The largest event known to have taken place in Texas occurred on August 16, 1931, near the town of Valentine. This magnitude 6.0 earthquake was associated with a number of foreshocks and aftershocks and produced damage to buildings over an area of several thousand square kilometers. A maximum Modified Mercalli intensity of VIII was reported. Earthquakes in several areas of Texas have also been attributed to oil-field operations. For example, in the Cogdell oil field, earthquakes may have been induced by fluid injection for secondary recovery operations or by withdrawal of fluids (Davis and Pennington, 1989).

## 12

Conclusions and speculations

A number of common motifs are intertwined in the myths and legends of earth-quakes of the Americas and are worth repeating: gods of the underworld, death, and the creation and end of the current and past worlds. North American Indians speak of the earthquake gods *Chief of the Below World, Yewol*, and *Cipas*, who are capable of shaking the ground when angry. The Lacandon Maya describe tales of *Sukukyum* and his brother-in-law *Cisin*, both of whom can trigger earthquakes. Cisin may also be the familiar God A of death referred to in the pre-conquest codices. *Chicchans* and *Cuch Uinahel Balumil* are the underground earthquake instigators of the Chorti and Tzotzil Maya. The stories may differ in detail with their abracadabra but there can be no doubt that earthquakes were viewed as punishment given from angry and restless residents of the underworld. Such tales and lore are a manifestation of how North and Mesoamerican natives viewed their world. They not only entertain us by watering the aridity of earthquake cataloging but also inform us that, for better or worse, earthquakes were a fabric of early life in the Americas. For example, belief in matters being animate and inanimate would not have made sense to the Maya. This cannot be better exemplified than the belief of the Quiché Maya of the Guatemalan high-lands that their life-sustaining maize is frightened of earthquakes; as a result the Quiché Maya often call to the maize during an earthquake to reassure it. Stated in another way, earthquake myths of the Americas represented the means for indigenous people to combine outer and inner world experiences and deal with the fragile network of everyday reality in the natural environment.

The Zuñi Indians of the American southwest, who reached their final place of settlement around AD 1350 at Itiwana, "the middle place", were said to have experienced earthquakes in their eastward migration. Whenever earthquakes occurred they regarded their place of settlement as unstable and moved to another.

Details of their migration may well be allegorical but they have certain core elements of truth. Earthquake occurrences, by themselves or associated with volcanic eruptions, are known, even today, to be a motivator of human mobility. In AD 1064 a major volcanic eruption took place just north of Flagstaff, Arizona, forming Sunset Crater. A 2000 km$^2$ area was covered by volcanic ash and this may have been a factor in the eastward migration of the Zuñi. Another plausible possibility for earthquakes triggering migration is that the seismic disturbances may have suppressed the flow of water from springs, forcing movement to more amenable surroundings.

The written historical earthquake record of the Americas does not pre-date the time of the Spanish conquest. At best the record is spatially and temporally incomplete and as a result it is not possible to discuss meaningfully concepts such as time-dependent estimates of future earthquakes. Such estimates are based on a very incomplete knowledge of past earthquake chronology. A fundamental premise of the movement of plate boundaries, which move at the rate of centimeters per year, is that the recurrence interval for significant earthquakes should be relatively short of the order of 100 to 400 years or so. With a historical record that is chronologically limited to a time scale of hundreds of years we have much to learn about "major earthquake cycles." It is clear that archaeoseismic and paleoseismic data can provide information to extend the chronology backward for several thousand years.

The primary data that are needed to determine fault slip rates and recurrence rates are the amount of offset and the age of the displaced features. Geologists traditionally carry out paleoseismic studies within a geological or stratigraphic framework. Piercing features, such as ancient shallow-buried stream channel deposits or marine terrace levels, are used to assess fault displacements. Estimated ages are obtained using radiocarbon analyses of carbon-bearing materials such as charcoal. In many cases, however, these techniques often produce results with large uncertainties and hence unknown recurrence intervals.

The use of archaeology to study past earthquake occurrences has not been the primary focus of archaeologists, although it is clear that ancient sites often contain ruins and features that can be examined and dated in well-defined and dated contexts for determining historical earthquake activity. There are great opportunities for collaboration between archaeologists and earth scientists in ancient earthquake studies. Two types of earthquake-induced damage need to be searched for in an archaeological expedition: (a) damage that is the result of co-seismic shaking and (b) features associated with surface breaks or ruptures.

Ancient societies that lived in zones subjected to earthquakes were undoubtedly as plagued by seismic events as we are today. These events can produce a visible impact and often leave spectacular episodes of destruction. Such external effects were not only adventitious but had to be disruptive. Certainly one cannot argue against the fact that many regions of the Americas occupied by indigenous people, such as the Maya and the Aztecs, have been subjected to earthquake and volcanic activity. Many explanations have been suggested for the abandonment of Maya sites in the mid ninth century. Unfortunately, Ockham's statement that "entities are not to be multiplied beyond necessity," is not in vogue. No single theory can uniformly answer all questions concerning the abandonment of Maya sites in the southern lowlands with a simple yes or no. The decline and last fitful gasp of this brilliant civilization spanned a period of military bellicosity and severe economic and environmental degradation. Individual cities may have been abandoned for a variety of reasons – climatic changes, earthquakes, epidemics, foreign and civil wars, cultural and social decay, agricultural and economic collapse. Whatever reason is argued for and accepted for the final collapse, a thesis is offered that earthquakes may have played a role in a scenario that led to the abandonment of individual sites. Earthquakes do not have to be great in size to produce shaking severe enough to damage monuments, stelae, walls, canals, and other structures. Perhaps continuing earthquake effects were the *coup de grâce* or provided the disincentive for continued rebuilding and reconstruction.

It was demonstrated that the sites of Quiriguá, Benque Viejo, and Copán could have been struck by an earthquake with an epicenter within the Caribbean plate boundary zone at the close of the Maya Classic Period. Other lowland sites such as Seibal, Altar de Sacrificios, and Pusilhá were found with stelae fallen in preferential directions suggestive of earthquake damage.

One of the arguments against earthquake damage in an archaeological context is that it is difficult to recognize earthquake effects because they might be obscured as a result of "poor construction and adverse geotechnical effects." It is further argued that damage can easily be ascribed to poor ground conditions or to earth movements that are unrelated to earthquakes, such as ground stresses imposed by construction of the site features themselves. These arguments are conceptually weak when considered in the context of the tectonic setting in which the communities were built. Inhabitants of sites located in or near zones of earthquake activity would certainly have perceived some danger to simple constructions, such as walls, from damaging levels of seismic intensity. Repeated rebuilding and patching of simple adobe walls and *tapiales* is common even today in the earthquake zones of Mexico, Central and South America. Evidence for strengthening of adobe walls by buttressing can be found at several Maya sites such as Quiriguá and Holmul, Guatemala. Clearly such simple buttressing can be viewed as an anti-earthquake response and suggests that the inhabitants were indeed subjected to continuing earthquake activity.

**Figure 12.1** Earthquake damage, from the December, 1934 earthquake, to a passageway in the east room of Structure 22 at Copán (Trik, 1939; courtesy of the Carnegie Institution of Washington).

Damage from recent earthquakes that took place at sites undergoing archaeological renovation suggests that the possibility of damage from earlier earthquakes cannot be ignored. Much building collapse is often obscured beneath a thick veneer of a growth of trees and vegetation (Figure 12.1).

More subtle evidence pointing to adaptability to earthquakes can be found upon examination of many Maya ruins. Flying façades and roof-combs are more prevalent in those areas less subjected to frequent earthquake activity. Roof-combs are not present at Quiriguá and Copán, sites located in the seismically active zone of the Caribbean plate boundary. The corbeled arch is not found in the earthquake-prone highland sites of Chiapas and Guatemala. A more critical observation of Maya architecture, however, is that the structures themselves possess the two key ingredients, symmetry and regularity, necessary for earthquake-resistant design.

Toppled monuments, gravestones, and stelae are often indications of past earthquake activity. A single fallen stela by itself, however, cannot be used to prove earthquake damage. It is the totality of the architectural scene and the prevalent common directions of fall for many stelae, together with an

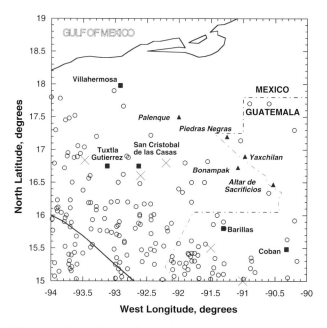

Figure 12.2 Seismicity in the Mexico–Guatemala area shown from January 1, 1975 to June 30, 2000. Events in the magnitude range $4.5 \geq M \leq 7.0$ are shown by open circles. X indicates events of $M \geq 7.1$ from 1816 to present.

appropriate and plausible level of seismic intensity, that argues for a probable earthquake cause. It seems unlikely that falling trees or vindictive individuals would systematically topple monuments in consistent directions.

The location of archaeological remains near past and present earthquake faults is certainly not random. Faults and fracture systems produce topographic features such as linear valleys and can also be a factor in controlling the alignment of cenotes, streams, lakes, and swales. Uplift and subsidence features can cause the local disruption of rivers. All of these geological features and landforms produce hydrological conditions that support the fauna and flora that are essential for human occupation. Surely indigenous cultures made use of the complex topography in tectonically active areas. Nevertheless, drainage basins and hydrological environments can change with time. It is not difficult to imagine that the subsequent abandonment of a site could have been the result of the earthquake deformation of the hydrological environment.

One of the criteria that should be used in assessing the likelihood of seismic damage to archaeological sites located in regions subjected to earthquakes is the level of shaking intensity that may have taken place over a specified area. If we adopt a Modified Mercalli intensity threshold value of VII to VIII it is empirically straightforward to estimate the potential area of strong shaking and significant damage, particularly for masonry and adobe structures. Earthquakes

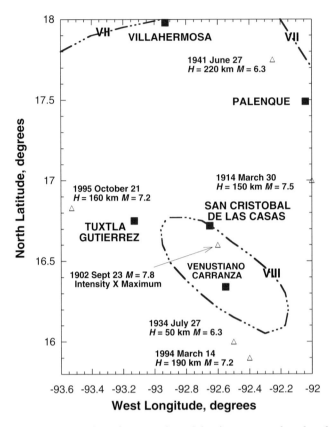

**Figure 12.3** Location of some major subduction-zone earthquakes that occurred in the state of Chiapas, Mexico. Isoseismals for the event of September 23, 1902 are shown. The epicenter of the event is uncertain. $H$ is the earthquake focal depth in kilometers.

with magnitudes of at least 7.5 occur on the average every 4 to 5 years somewhere along the Pacific rim of the Americas. An earthquake of this magnitude would produce an area of $30\,000$ km$^2$ that was subjected to intensity VII or greater. With an area this size it is easily seen that earthquakes have affected large regions of Mexico, Central America, Peru, and Chile. For example, the historical record for Peru shows that since 1582 there have been 20 great subduction-zone earthquakes that took place within the south latitude range of $6°$ to $20°$. As a result, a minimum area of approximately $280\,000$ km$^2$ of the coastal and Andean foothill region of Peru has been subjected to a Modified Mercalli intensity value of at least VII to VIII or greater.

Excavations at the nuclear America site of Kotosh ($9.9°$ S, $76.3°$ W), near the modern Peruvian city of Huánuco on the eastern slope of the Andes, have exposed sections of at least five different stratigraphic epochs of stone

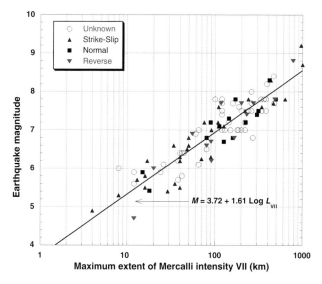

**Figure 12.4** Length of Mercalli intensity VII versus magnitude. Data are tabulated in Appendix A. Symbols refer to the type of earthquake faulting mechanism: open circles, unknown; triangles, strike-slip mechanisms; squares, normal faulting; inverted triangles, reverse mechanisms.

constructions demonstrating a rich archaeological history. Were the constructions buried with earth by human hands after periods of abandonment or can other reasons be found for the repeated reconstructions? No conclusions have been drawn as to why Kotosh went to rack and ruin. There is obvious room for further examination of its natural environment and critical assessment of whether earthquakes were the reason for repeated rebuilding.

In the search for evidence for ancient earthquake damage, the present-day condition of architectural ruins needs to be considered. What is seen today is not necessarily the condition in which many ancient structures were found. For example, the great Classic Maya city of Palenque built in Chiapas, Mexico, lies in a region periodically subjected to earthquake activity (Figure 12.2). Can evidence for earthquake damage be found in the ruins? Palenque, in the foothills of the Chiapas mountains, was active from AD 600–900, reaching its peak of artistic achievement in AD 692. After its abandonment the jungle overtook Palenque and created a veil of oblivion although native Indians are known to have lived in its vicinity since ancient times. The ruins of Palenque first became known to the Spaniards in 1773.

Palenque is located in a geographical region that is subjected to the effects of earthquakes produced by the thinner subducting plate moving northeasterly from the Pacific Ocean beneath the Mexican isthmus. The earthquake event

**Figure 12.5** Photograph of the Great Palace of Palenque taken in 1891 by Maudslay (1889–1902). View is from the southwest. Note that vegetation appears to be supporting the square tower of the Great Palace.

locations shown in Figures 12.2 and 12.3 are taken from catalogs maintained by the National Earthquake Information Center (http://www.neic.cr.usgs.gov/neis/epic/epic.html) and the International Seismological Centre (http://www.isc.ac.uk). Although earthquakes of magnitude greater than 7 may occur less frequently in the Chiapas region of Mexico compared to other parts of coastal Mexico, and are of intermediate focal depths, they are capable of producing damage over wide areas.

Figure 12.4 shows a plot of the length of intensity VII on the Modified Mercalli Scale as a function of magnitude for many earthquakes from a wide variety of tectonic settings. Even with the range of intensity values exhibited by earthquakes of a specific magnitude from diverse tectonic and geographic settings it is clear that earthquakes of $M > 7$ can affect large areas. For example, the event of September 23, 1902, with an estimated magnitude of at least 7.8, was described as the most disastrous earthquake of the twentieth century experienced in the states of Chiapas and Tabasco. Intensities of VI to X on the Modified Mercalli Scale were reported. Surely, events of such a size cannot be summarily precluded from having struck Palenque in the past.

**Figure 12.6** Semi-collapsed church tower in Puebla, Mexico, damaged in the $M_w = 6.7$ Tehuacán earthquake of June 15, 1999. The Mercalli intensity level in Puebla was VII (Jiménez *et al.*, 1999).

It is not surprising that the explorers of the nineteenth century would not have considered the possibility of any historical earthquake damage at Palenque. The mounds on which the buildings stood and their peripheral areas were covered with fallen stones, debris, and vegetation, and had been ruthlessly subjected to tree leveling and burning by early explorers (Figure 12.5). Many of the ornamented superstructures had almost entirely disappeared. However, towers and other tall structures or appendages are particularly vulnerable to earthquake shaking. Damage is primarily from shearing in the areas around openings. Diagonal cracking and in some instances partial or total collapse of the tower takes place. An example of a tower severely damaged by earthquake shaking is shown in Figure 12.6. A structure that has probably been subjected to past

(a)

(b)

(c)

(d)

**Figure 12.7** Reconstruction, beginning in 1949, of the upper part of the tower of the Great Palace of Palenque, Mexico, to its present state. Photograph (a) looks south, all other views look northeasterly (photographs (a), (b) and (c) are from Lhuillier [1952a, b, c]; photograph (d) is by the author).

earthquake damage is the square tower of the Great Palace of Palenque and its adjoining cresterias. It is important to point out that this tower has been carefully *restored* to its present impressive state. The sequence of photographs shown in Figure 12.7 demonstrates the reconstruction of the tower at Palenque. The possibility of past earthquake damage to the tower at Palenque is intriguing.

The past 400 years of history are full of written accounts of seismic events that have produced tragic and often catastrophic consequences. Archaeology provides the tool for extending the historical record of earthquakes further back in time. The evidence for earthquake damage is often only suggestive but is bolstered by consistency arguments and the *onus probandi* is necessarily placed upon those who wish to prove one point or another. The collapse of structures and sequential rebuilding using different types of construction are important chronologic horizons offering clues to earthquake occurrences in areas of long-term human occupation. Such chronologic information can often be placed in a cultural and evolutionary framework. There is much yet to be uncovered.

# Appendix A Data relating magnitude to extent of Modified Mercalli intensity of VII

| Date | Location | Mechanism | Maximum extent of MMI = VII (km) | Magnitude | Reference |
|---|---|---|---|---|---|
| July 22, 1816 | Guatemala | Strike-slip | 340 | 7.6 | White (1985) |
| January 1, 1837 | Safed, Israel | Strike-slip | 90 | 6.3 | Vered and Striem (1977) |
| January 9, 1857 | Fort Tejon, California | Strike-slip | 460 | 7.7 | Agnew and Sieh (1978) |
| March 26, 1872 | Owens Valley, California | Normal | 416 | 8.3 | Greensfelder (1972) |
| June 5, 1897 | Tehuantepec, Mexico | Unknown | 200 | 7.0 | Figueroa (1963b) |
| June 12, 1897 | Assam, India | Strike-slip | 1000 | 8.7 | Oldham (1899) |
| January 16, 1902 | Mexico | Unknown | 130 | 7.0 | Figueroa (1963b) |
| September 23, 1902 | Chiapas, Mexico | Unknown | 360 | 7.8 | Figueroa (1963b) |
| April 4, 1904 | Bulgaria | Unknown | 470 | 7.7 | Papazachos et al. (1997) |
| January 31, 1906 | Colombia | Reverse | 778 | 8.8 | Kanamori and McNally (1992) |
| April 18, 1906 | San Francisco, California | Strike-slip | 630 | 7.8 | Lawson et al. (1908) |
| August 16, 1906 | Valparaíso | Strike-slip | 430 | 8.3 | Ballore (1912) |
| March 26, 1908 | Mexico | Unknown | 260 | 7.5 | Figueroa (1963b) |
| July 30, 1909 | Mexico | Unknown | 160 | 7.7 | Figueroa (1963b) |
| July 31, 1909 | Mexico | Unknown | 120 | 7.0 | Figueroa (1963b) |

*(cont.)*

| Date | Location | Mechanism | Maximum extent of MMI = VII (km) | Magnitude | Reference |
|------|----------|-----------|----------------------------------|-----------|-----------|
| October 31, 1909 | Mexico | Unknown | 130 | 7.0 | Figueroa (1963b) |
| June 7, 1911 | Mexico | Unknown | 260 | 8.0 | Figueroa (1963b) |
| December 16, 1911 | Mexico | Unknown | 220 | 7.0 | Figueroa (1963b) |
| July 19, 1912 | El Salvador | Unknown | 24 | 5.9 | White and Harlow (1993) |
| June 23, 1915 | Imperial Valley, California | Strike-slip | 40 | 6.2 | Beal (1915) |
| October 3, 1915 | Pleasant Valley, Nevada | Normal | 171 | 7.8 | Jones (1915) |
| December 29, 1915 | Honduras | Unknown | 40 | 6.4 | White and Harlow (1993) |
| June 8, 1917 | El Salvador | Unknown | 42 | 6.4 | White and Harlow (1993) |
| December 26, 1917 | Guatemala | Unknown | 42 | 5.8 | White and Harlow (1993) |
| January 3, 1920 | Mexico | Unknown | 120 | 7.8 | Figueroa (1963b) |
| June 21, 1920 | Inglewood, California | Strike-slip | 4 | 4.9 | Richter (1970) |
| June 29, 1925 | Santa Barbara, California | Reverse | 91 | 6.2 | Symposium (1925) |
| November 2, 1927 | Jericho, Israel | Strike-slip | 70 | 6.3 | Vered and Striem (1977) |
| February 9, 1928 | Oaxaca, Mexico | Unknown | 120 | 7.7 | Figueroa (1963b) |
| March 21, 1928 | Mexico | Unknown | 240 | 7.5 | Figueroa (1963b) |
| April 16, 1928 | Puebla, Mexico | Unknown | 260 | 7.7 | Figueroa (1963b) |
| April 18, 1928 | Bulgaria | Unknown | 185 | 7.0 | Papazachos *et al.* (1997) |
| June 17, 1928 | Mexico | Unknown | 120 | 7.5 | Figueroa (1963b) |
| August 4, 1928 | Mexico | Unknown | 120 | 7.4 | Figueroa (1963b) |
| June 16, 1929 | Buller, New Zealand | Unknown | 333 | 7.8 | Downes (1995) |
| June 16, 1929 | Murchison, New Zealand | Reverse | 480 | 7.8 | Anderson *et al.* (1994) |
| July 28, 1929 | Whittier, California | Reverse | 12 | 4.7 | Richter (1958a) |

*(cont.)*

(*cont.*)

| Date | Location | Mechanism | Maximum extent of MMI = VII (km) | Magnitude | Reference |
|---|---|---|---|---|---|
| July 14, 1930 | Guatemala | Unknown | 60 | 6.9 | White and Harlow (1993) |
| January 15, 1931 | Oaxaca, Mexico | Normal | 489 | 7.8 | Singh *et al.* (2000) |
| June 3, 1932 | Jalisco, Mexico | Unknown | 388 | 8.2 | Singh *et al.* (1985) |
| December 21, 1932 | Nevada | Normal | 144 | 7.3 | Gianella and Callaghan (1934) |
| January 15, 1934 | Bihar-Nepal | Unknown | 461 | 8.4 | Officers of the Geological Survey (1939) |
| December 3, 1934 | Copán, Honduras | Unknown | 38 | 6.2 | White and Harlow (1993) |
| June 24, 1942 | Wairarapa, New Zealand | Unknown | 144 | 7.2 | Downes (1995) |
| November 10, 1946 | Ancash, Peru | Normal | 89 | 7.2 | Silgado (1951) |
| December 4, 1948 | Desert Hot Springs, California | Strike-slip | 128 | 6.5 | Richter *et al.* (1958) |
| May 21, 1950 | Cusco, Peru | Unknown | 8 | 6.0 | Ericksen *et al.* (1954) |
| December 14, 1950 | California | Unknown | 12 | 5.6 | Murphy and Ulrich (1952) |
| July 21, 1952 | Kern County, California | Reverse | 272 | 7.7 | Murphy and Cloud (1954) |
| June 14, 1953 | California | Strike-slip | 16 | 5.5 | Murphy and Cloud (1955) |
| January 12, 1954 | California | Unknown | 12 | 5.9 | Murphy and Cloud (1956) |
| April 25, 1954 | California | Strike-slip | 8 | 5.3 | Murphy and Cloud (1956) |
| July 6, 1954 | Nevada | Normal | 80 | 6.8 | Murphy and Cloud (1956) |
| August 24, 1954 | Nevada | Normal | 160 | 6.8 | Murphy and Cloud (1956) |
| December 16, 1954 | Nevada | Normal | 224 | 7.2 | Murphy and Cloud (1956) |
| December 21, 1954 | California | Strike-slip | 48 | 6.5 | Murphy and Cloud (1956) |

(*cont.*)

| Date | Location | Mechanism | Maximum extent of MMI = VII (km) | Magnitude | Reference |
|---|---|---|---|---|---|
| September 5, 1955 | California | Strike-slip | 16 | 5.8 | Murphy and Cloud (1957) |
| October 24, 1955 | California | Strike-slip | 16 | 5.8 | Murphy and Cloud (1957) |
| April 25, 1957 | Rhodos, Greece | Unknown | 220 | 7.2 | Papazachos *et al.* (1997) |
| July 28, 1957 | Mexico | Unknown | 200 | 7.5 | Figueroa (1963b) |
| May 24, 1959 | Mexico | Unknown | 270 | 6.8 | Figueroa (1963b) |
| August 18, 1959 | Hebgen Lake, Montana | Normal | 112 | 7.1 | Eppley and Cloud (1961) |
| August 26, 1959 | Mexico | Unknown | 80 | 6.5 | Figueroa (1963b) |
| April 9, 1961 | California | Strike-slip | 35 | 5.6 | Lander and Cloud (1963) |
| September 1, 1962 | Buyin-Zara, Iran | Strike-slip | 108 | 7.2 | Ambraseys (1963) |
| July 26, 1963 | Greece | Unknown | 65 | 6.1 | Zátopek (1968) |
| June 27, 1966 | Parkfield, California | Strike-slip | 40 | 5.5 | Cloud (1967) |
| August 19, 1966 | Varto, Turkey | Strike-slip | 50 | 6.8 | Ambraseys and Zátopek (1968) |
| October 17, 1966 | Peru | Unknown | 340 | 7.5 | Lomnitz and Cabré (1968) |
| March 27, 1964 | Alaska | Strike-slip | 960 | 9.2 | von Hake and Cloud (1966) |
| July 22, 1967 | Mudurnu, Turkey | Strike-slip | 91 | 7.0 | Ambraseys and Zatopek (1969) |
| April 9, 1968 | Borrego Mountain, California | Strike-slip | 50 | 6.6 | Hanks *et al.* (1975) |
| May 23, 1968 | Inanagahua, New Zealand | Reverse | 230 | 7.4 | Anderson *et al.* (1994) |
| August 31, 1968 | Dasht-e Bayaz, Iran | Strike-slip | 126 | 7.1 | Ambraseys and Tchalenko (1969) |
| October 14, 1968 | Meckering, Australia | Reverse | 55 | 6.9 | Gordon and Lewis (1980) |
| October 2, 1969 | Santa Rosa, California | Strike-slip | 13 | 5.7 | Steinbrugge *et al.* (1970) |

(*cont.*)

(*cont.*)

| Date | Location | Mechanism | Maximum extent of MMI = VII (km) | Magnitude | Reference |
|---|---|---|---|---|---|
| March 28, 1970 | Gediz, Turkey | Strike-slip | 110 | 7.1 | Tasdemiroglu (1971) |
| May 31, 1970 | Peru | Unknown | 380 | 7.8 | Gajardo (1970) |
| February 9, 1971 | San Fernando, California | Reverse | 80 | 6.6 | Coffman and von Hake (1973b) |
| July 8, 1971 | Chile | Unknown | 380 | 7.5 | Eisenberg *et al.* (1972) |
| October 12, 1971 | Barillas, Guatemala | Unknown | 38 | 5.7 | White and Harlow (1993) |
| December 23, 1972 | Managua, Nicaragua | Strike-slip | 17 | 6.2 | Hansen and Chavez (1973) |
| February 21, 1973 | Pt. Mugu, California | Reverse | 20 | 6.0 | Coffman and von Hake (1975) |
| October 3, 1974 | Lima, Peru | Unknown | 188 | 7.6 | Espinosa *et al.* (1977) |
| February 4, 1976 | Guatemala | Strike-slip | 220 | 7.5 | Espinosa (1976) |
| July 27, 1976 | China | Strike-slip | 240 | 7.8 | Yong *et al.* (1988) |
| March 21, 1977 | Bandar Abbas, Iran | Unknown | 65 | 7.0 | Berberian and Papastamatiou (1978) |
| September 16, 1978 | Tabas-e-Golsham, Iran | Reverse | 117 | 7.7 | Berberian (1979) |
| October 15, 1979 | Imperial Valley, California | Strike-slip | 40 | 6.3 | Nason (1982) |
| November 14, 1979 | Ghaenat, Iran | Strike-slip | 34 | 6.6 | Haghipour and Amidi (1980) |
| November 27, 1979 | Ghaenat, Iran | Strike-slip | 75 | 7.2 | Haghipour and Amidi (1980) |
| August 9, 1980 | Guatemala | Unknown | 44 | 6.4 | White and Harlow (1993) |
| February 24, 1981 | Greece | Normal | 126 | 6.7 | Carydis *et al.* (1982) |
| April 5, 1986 | Cuzco, Peru | Normal | 18 | 5.4 | Cabrera and Sébrier (1998) |
| September 18, 1991 | Guatemala | Unknown | 26 | 6.1 | White and Harlow (1993) |
| June 28, 1992 | Landers, California | Strike-slip | 110 | 7.6 | Brewer (1992) |

(*cont.*)

| Date | Location | Mechanism | Maximum extent of MMI = VII (km) | Magnitude | Reference |
|------|----------|-----------|----------------------------------|-----------|-----------|
| January 17, 1994 | Northridge, California | Reverse | 90 | 6.7 | Dewey (1994) |
| February 3, 1994 | Idaho | Normal | 15 | 5.9 | Schuster and Murphy (1996) |
| October 3, 1995 | Ecuador | Reverse | 166 | 6.8 | Yepes *et al.* (1996) |
| November 12, 1996 | Nasca, Peru | Reverse | 200 | 7.7 | Chatelain *et al.* (1997) |
| June 15, 1999 | Tehuacán, Mexico | Unknown | 180 | 7.0 | Jiménez *et al.* (1999) |
| September 30, 1999 | Oaxaca, Mexico | Normal | 311 | 7.5 | Singh *et al.* (2000) |
| October 16, 1999 | Hector Mine, California | Strike-slip | 110 | 7.1 | Anon. (2000) |

The line shown in Figure 12.4 is a log-linear regression line between earthquake magnitude and the maximum dimension of the area enclosed by the Modified Mercalli intensity of VII for 106 events. The functional form $M = 3.72 + 1.61 \log L_{VII}$ for all- slip-type earthquakes has a correlation coefficient $r = 0.91$. If only dip-slip events are considered (27 normal and reverse) the functional form is $M = 3.48 + 1.70 \log L_{VII}$. The magnitude difference for a given $L_{VII}$ does not exceed 0.2 magnitude units, arguing that the regression shown for all slip types is appropriate for this type of estimation.

# Appendix B  Data relating magnitude to area of Mexican earthquakes with Modified Mercalli intensity ≥ VI

| Date | Latitude (° N) | Longitude (° W) | Area (km²) | Magnitude | Reference |
|------|---------------|-----------------|-----------|-----------|-----------|
| May 3, 1887 | 30.80 | 109.25 | 353 250 | 7.4 | DuBois and Smith (1980) |
| September 23, 1902 | 16.00 | 93.00 | 162 000 | 7.8 | Singh *et al.* (1980) |
| March 26, 1908 | 16.70 | 99.20 | 85 800 | 7.8 | " |
| July 30, 1909 | 16.80 | 99.90 | 60 800 | 7.4 | " |
| June 7, 1911 | 19.70 | 103.70 | 94 200 | 7.7 | " |
| August 27, 1911 | 16.77 | 95.90 | 58 600 | 6.75 | " |
| December 16, 1911 | 16.90 | 100.70 | 60 700 | 7.5 | " |
| February 10, 1928 | 18.26 | 97.99 | 19 700 | 6.5 | " |
| March 22, 1928 | 16.00 | 96.00 | 105 200 | 7.5 | " |
| June 17, 1928 | 16.30 | 96.70 | 39 400 | 7.9 | " |
| August 4, 1928 | 16.83 | 97.61 | 36 200 | 7.4 | " |
| January 15, 1931 | 16.10 | 96.64 | 94 200 | 7.8 | Singh *et al.* (2000) |
| April 15, 1941 | 18.85 | 102.94 | 218 500 | 7.7 | Singh *et al.* (1980) |
| November 9, 1956 | 17.00 | 94.00 | 67 100 | 6.4 | " |
| July 28, 1957 | 17.11 | 99.10 | 224 500 | 7.5 | " |
| May 24, 1959 | 17.61 | 97.17 | 91 000 | 6.9 | " |
| August 26, 1959 | 18.26 | 94.43 | 23 000 | 6.75 | " |
| May 11, 1962 | 17.25 | 99.58 | 130 400 | 7.0 | " |
| May 19, 1962 | 17.12 | 99.57 | 164 900 | 7.2 | " |
| August 23, 1965 | 16.30 | 95.80 | 146 200 | 7.6 | " |
| August 2, 1968 | 16.60 | 97.70 | 69 200 | 7.4 | " |
| January 30, 1973 | 18.39 | 103.21 | 222 700 | 7.5 | " |
| August 28, 1973 | 18.30 | 96.60 | 114 000 | 7.2 | " |
| November 29, 1978 | 16.01 | 96.59 | 57 000 | 7.7 | Singh *et al.* (1981) |
| March 14, 1979 | 17.31 | 101.35 | 225 400 | 7.6 | Singh *et al.* (1980) |
| October 24, 1980 | 18.21 | 98.24 | 82 300 | 7.0 | Singh *et al.* (1981) |
| June 15, 1999 | 18.39 | 97.44 | 58 150 | 7.0 | Singh *et al.* (1999) |
| September 30, 1999 | 16.06 | 96.93 | 62 800 | 7.5 | Singh *et al.* (2000) |

The heavy solid line shown in Figure B.1 is a correlation by Toppozada (1975) for California and Nevada earthquakes. The data for Mexican earthquakes systematically show smaller Modified Mercalli intensity VI areas compared to California and Nevada earthquakes. Isoseismal areas in Mexico, particularly for events of the early twentieth century, may have been underestimated because of the sparseness of population density or may represent geographical differences in the attenuation of ground motion.

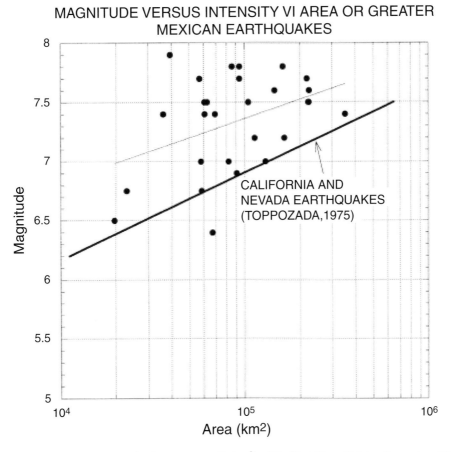

**Figure B.1** Magnitude versus area (in km$^2$) of Modified Mercalli intensity contour VI. Estimate for California earthquakes is shown for comparison.

# Appendix C   Rossi–Forel Seismic Intensity Scale

The Rossi–Forel Seismic Intensity Scale (RF) was developed in 1883 for use in studying Italian earthquakes. It uses a roman numeral scale of I to X and describes earthquake intensity in terms of observed wall-cracking effects and levels of shaking experienced by people. The scale and the comparison of its values with the Modified Mercalli Intensity Scale (MMI) are shown in Table C.1.

Table C.1. *Rossi–Forel Seismic Intensity Scale*

| Rossi–Forel value | Description |
| --- | --- |
| I | Recorded by a single seismograph or by seismographs of the same model but not by several seismographs of different kinds; the shock felt by an experienced observer. |
| II | Recorded by several seismographs of different kinds; felt by a small number of persons at rest. |
| III | Felt by several persons at rest; strong enough for the duration or direction to be appreciable. |
| IV | Felt by several persons in motion; disturbance of moveable objects, doors, windows, creaking of floors, cracking of ceilings. |
| V | Felt generally by everyone; disturbance of furniture and beds; ringing of some bells. |
| VI | General awakening of those asleep; general ringing of bells; oscillation of chandeliers; stopping of clocks; visible agitation of trees and shrubs; some startled persons leave their dwellings. |
| VII | Overthrow of movable objects; fall of plaster; ringing of church bells; general panic, without serious damage to buildings. |
| VIII | Fall of chimneys; cracks in the walls of buildings. |
| IX | Partial or total destruction of some buildings. |
| X | Great disaster; ruins; disturbance of strata; fissures in the ground; rock-falls from mountains; landslides. |

Table C.2. *Comparison of RF and MMI*

| MMI | I | II | III | IV | V |
|---|---|---|---|---|---|
| RF | I | I–II | III | IV–V | V–VI |
| MMI | VI | VII | VIII | IX | X–XII |
| RF | VI–VII | VIII– | VIII+ to IX– | IX+ | X |

The correspondence of the Rossi–Forel and the Modified Mercalli Intensity Scale is shown in Table C.2.

The correlation is not completely satisfactory. High intensities using the Modified Mercalli Scale are grouped together and the comparison of the middle grades is equivocal if only a published map is available. One should not expect to see the effects of intensity levels of less than VII preserved in archaeological ruins.

# Glossary

**Adobe** A sun-dried clay brick.

**Aseismic slip** Movement along a fault zone that is not accompanied by an earthquake.

**Baktun** Mayan chronologic unit of time equal to 144 000 days or 394.5 years.

**Batter** A gradual upward and backward slope on a wall.

**Buttress** A projecting structure, usually of stone, built against a wall to strengthen or reinforce it. A common anti-seismic construction technique.

**Caban** Seventeenth day of the 20-day Maya month. Word means earth. Believed by some scholars to be the earthquake sign.

**Cabrakán (earthquake)** Son of Vukub-Cakix, the great macaw, described in the Quiché *Popul Vuh*.

**Cakchiquel** Ethnic Maya group found in the western highlands of Guatemala.

**Cap stone** The stone that closes the false arch or Maya vault.

**Causative fault** A term used to designate a fault producing a significant earthquake.

**Characteristic earthquake** An earthquake in which slip at any point along the fault is the same from one earthquake to the next.

**Chilam Balams** Pre-conquest- and Colonial-era Maya manuscripts of the northern Yucatan describing local history. Translated into phonetic Maya and Spanish in Colonial times.

**Chorti** Ethnic Maya group and language from southeastern Guatemala and western Honduras.

**Classic Period** A designation in Maya archaeology for the time period AD 300–900. The fluorescent period of Tikal, Palenque, and Copán.

**Codex** Pre-Hispanic manuscripts on leather or treated bark, as well as Colonial ones on paper or cloth of European origin. The plural is "codices".

**Colluvium** Alluvium deposited by surface run-off, usually at the base of a slope.

**Corbel vault** Arches built up horizontally, unlike true vaulting, which makes use of voussoirs and a keystone.

**Cresteria** Spanish name given to the roof-comb on Maya buildings.

**Dip-slip fault** A fault in which the offset or movement is along the direction of the dip of the fault. Dip is measured from the horizontal. Fault offset can be either normal or reverse.

**Drift (of a structure)** The maximum deflection from the vertical of the top of a structure, induced by ground shaking.

**Epicenter** The point on the Earth's surface directly above the focus of an earthquake.

**Glyph** Term, analogous to Egyptian hieroglyphs, referring to the complex symbols of Maya and later Aztec writing. Glifo in Spanish.

**Graben** A block of rock, commonly long and narrow, that has dropped down along boundary faults.

**Ground acceleration** The rate of increase of speed or velocity. Ground accelerations due to earthquake motions are measured relative to 980 cm/s$^2$ or 1 g.

**Holocene** 10 000 years before the present.

**Intensity (of earthquakes)** A numerical assessment of ground shaking obtained from observations of damage to structures, changes in the ground surface, and felt reports.

**Intraplate earthquake** Earthquake with its epicenter or focus within a plate.

**Katun** Period of 7200 days in the Mayan calendar, or slightly less than 20 years.

**Kin** A day in the Mayan calendar.

**Lacandon** An ethnic Mayan group and language. The Lacandons inhabit the upper reaches of the Usumacinta river in the state of Chiapas, Mexico.

**Left-lateral fault** A strike-slip fault in which, regardless of from which side it is observed, the opposite side has moved to the left.

**Liquefaction** The process whereby a mixture of soil and sand behaves like a fluid during earthquake shaking.

**Long Count** System of Mayan date calculation using baktuns, katuns, tuns, uinals, and kins, starting from an initial date corresponding to 3113 BC.

**Magnitude (earthquake)** A measure of earthquake size determined by taking the logarithm to the base 10 of the largest ground motion of a seismic wave-type recorded on a seismograph and corrected for the distance to the epicenter. Three scales are in use: Richter or local magnitude, body-wave magnitude, and surface-wave magnitude, $M_s$.

**Magnitude (tsunami)** An empirical measure of the size of a tsunami based on the local height of the water wave.

**Meizoseismal area** The region of strong ground shaking and significant damage in an earthquake.

**Mesoamerica** The geographic area roughly situated between the United States and Costa Rica.

**Moment magnitude** Magnitude of an earthquake estimated from the seismic moment.

**Nahuatl** Ethnic group in Mexico speaking the Nahuatl tongue, including the Toltecs and the Aztecs. The lingua franca of Mesoamerica.

**Normal fault** A dip-slip fault in which the hanging wall block above the fault has moved downward relative to the footwall block.

*Ollin* Seventeenth of the 20 Nahuatl day signs signifying movement or earthquake. Corresponds to the Maya day *Caban* and the Zapotec day *Xòo*.

**Petén** The central lowland zone in northern Guatemala occupied by the Maya.

*Popul Vuh* Book written around 1557 in the Quiché language of the Maya, but using the Latin alphabet.

**Pre-Columbian (Pre-conquest, Pre-Cortés)** Designates civilizations that existed in America prior to arrival of Columbus. Pre-conquest designates the peoples of Mesoamerica before the conquest of Mexico by Cortés in 1522.

**Pyramid** In Mesoamerica a mound or raised platform constructed to support a temple. Not a true pyramid in the Egyptian sense.

**Pyroclastics** Fragmentary rocks ejected from a volcanic vent, often in great volume.

**Quiché** An ethnic language and group of the Maya found in the highlands of Guatemala.

**Radiation pattern** The pattern of pushes and pulls produced by seismic waves radiating from the seismic focus. It contains information as to the orientation of the causative fault.

**Repeat (recurrence) time** The interval time between earthquake occurrences of the same magnitude.

**Resonance** A large vibration when a structure is driven at the frequency special to that structure.

**Reverse fault** A dip-slip fault in which the hanging wall block moves upward over the footwall block.

**Roof-comb** A lofty decoration intended to heighten the appearance of a structure. It is often formed from two open-lattice walls leaning against each other with ornamental relief structures or key patterns. Sometimes called a flying façade if placed on the top-front of a structure giving the building a false front. Flying façades are not found in southern Maya cities.

**Run-up height (of tsunami)** The elevation reached by the level of water when a tsunami runs up onto the shoreline.

**Sand blow** The emission of fine-grained sediment in suspension with water as a result of violent earthquake shaking. May at times form small mounds known as sand volcanoes.

**Scarp** A cliff or near-vertical slope formed along a fault by ground movements during an earthquake.

**Seismic moment** A measure of earthquake size related to the leverage of the forces applied across an area of fault slip. Mathematically equivalent to the product of the rigidity of the rock times the area of faulting in the plane of rupture times the amount of slip. Units are dyne cm or Newton meters.

**Shaman** Native religious practitioner, the link between the living and the unseen world of Gods.

**Slickensides** Striations left on an escarpment face as a result of rocks sliding past each other during faulting.

**Slip rate** The rate of movement between plate boundaries. Can be measured geodetically or inferred from geophysical measurements.

**Stela** An upright stone slab, shaft, or pillar carved with an inscription or design. The plural is "stelae".

**Strike of a fault** The line of intersection between a fault plane and the surface of the ground. Its position is measured as an angle clockwise from north.

**Strike-slip fault** A fault whose relative offset or displacement is purely horizontal.

**Subduction zone** A dipping oceanic plate being forced beneath a thicker continental plate. It is the locus of intermediate- and deep-focus earthquakes defining the Benioff zone.

**Substructure** Construction covered by a later building.

**Tapiale** A large block formed by tapping mud into a frame. Used in the construction of walls.

**Talud-tablero** An architectural style that used a sloping lower wall surmounted by rectangular panels with a projecting framework.

**Terremoto** Spanish word for destructive earthquake. "Temblor" designates an earthquake of less severity.

**Tertiary** The division of geological time starting 65 million years ago and ending approximately 2 million years ago.

**Time-predictable earthquake model** An earthquake recurrence model based on the premise that the sum of individual slips over many earthquakes must agree with the overall plate tectonic slip rate.

**Tonalamatl** Mexican book of the days recording a ritual calendar of 260 days.

**Transform fault** A strike-slip fault. Pairs of plates slide past each other along transform faults.

**Tsunami** A large, long-wavelength water wave produced by a sudden change in the elevation of the sea-floor from an earthquake, a volcanic eruption or a submarine landslide.

**Tun** 360 days in the Mayan calendar.

**Tzotzil** Ethnic Maya group and language from the highlands of Chiapas, Mexico.

**Uinal** 20 days in the Mayan calendar. 18 uinals comprise a tun of 360 days.

**Vigesimal system** A positional system of numbers used by the Maya. Its base is the number 20. One changes columns after 19, 399, 7999, etc.

**Xibalba** Kingdom of the Mayan underworld according to the *Popol Vuh*.

**Zapotecs** Pre-Columbian people from the region of Oaxaca, Mexico. They flourished between AD 250 and 800.

# Bibliographic summaries

The study of earthquakes of the Americas has produced a broad and varied literature about seismic geography; myths and legends; hazards; Maya, Aztec, and Inca history; archaeological excavations and paleoseismicity. The literature is vast, ranging from the popular to the archival with scholarly interpretations and speculations. An attempt was made to be comprehensive in the bibliography but some relevant sources may have been inadvertently missed. The citations given here are organized by the chapters in which the text has made use of the opinions and ideas about the topics under discussion. Full citations of the sources used here are given in the Reference section that follows these chapter summaries. Bibliographies in the referenced publications themselves will provide the serious investigator with additional sources of information.

## Chapter 1 Introduction

*Early seismic catalogs* Gutenberg and Richter, 1941, 1949; Mallet and Mallet, 1858; Milne, 1912.

*Earthquake magnitude* Hanks and Kanamori, 1979; Kanamori, 1977, 1978, 1983; Kanamori and Anderson, 1975; Richter, 1935a, 1958, 1970; Singh and Havskov, 1980; Toppozada, 1975.

*Precarious rocks* Brune, 1996, 1999, 2002.

*Archaeology, stories, and earthquakes* Clements and Clements, 1953; Craine and Reindorp, 1979; Cushing, 1890; Edmonson, 1982; Haury, 1945; Housner, 1963; Hunt, 1960; Hunt, 1975; Kovach, 1998; Morley, 1920; Nelson, 1891; Noller *et al.*, 1994; Nur and Cline, 2000; Nur and Ron, 1997a, b; Sangawa, 1992; Spier, 1933; Taber, 1920b; Wallace and Taylor, 1955; Willey, 1966.

*Abandonment of Maya sites* Adams, 1973; Adams and Smith, 1977; Brunhouse, 1973, 1975; Cooke 1931; Cowgill, 1964; Culbert, 1973; Hammond, 1974, 1982, 1998; Hammond and Willey, 1979; Hodell *et al.*, 1995; Lowe, 1985; Sabloff, 1990; Thompson, 1954, 1970; Webster, 2002; Willey, 1987; Willey and Shimkin, 1971.

## Chapter 2 Earthquake mosaic of the Americas

*Zonal perspective* Abe and Kanamori, 1980; Anderson *et al.*, 1996; Bonilla *et al.*, 1984; Coffman and von Hake, 1973a, b; Davison, 1936; Ganse and Nelson, 1981;

Geller and Kanamori, 1977; Hobbs, 1907; Isacks *et al.*, 1968; Kanamori and Abe, 1979; Montandan, 1962; Montessus de Ballore, 1911; Pacheco and Sykes, 1992; Sieberg, 1932; Stover and Coffman, 1993; Talley and Cloud, 1962; Wells and Coppersmith, 1994; Yeats *et al.*, 1997.

**Northwest coastal Pacific zone**  Adams, 1990; Allen, 1975; Allen *et al.*, 1965; Anderson and Bodin, 1987; Atwater, 1987, 1992; Atwater and Moore, 1992; Atwater and Hemphill-Haley, 1997; Berg and Baker, 1963; Bolt and Miller, 1975; Bradford, 1935; Brown *et al.*, 1973; Bulletins of the Berkeley Seismographic Stations, 1973–92; Carpenter, 1921; Cassidy *et al.*, 1988; Cole *et al.*, 1996; Corbett and Johnson, 1982; Hauksson, 1987; Hauksson and Gross, 1991; Heaton and Kanamori, 1984; Heaton and Hartzell, 1987; Hileman *et al.*, 1973; Holden, 1892, 1898; Jacoby *et al.*, 1997; Kerr, 1995; Louderback, 1947; McAdie, 1907; Peterson and Wesnousky, 1944; Rasmussen, 1967; Rogers *et al.*, 1996; Satake *et al.*, 1996; Townley and Allen, 1939; Wood, 1916, 1933; Wood *et al.*, 1939.

**Western coastal zone**  Adamek *et al.*, 1988; Algermissen *et al.*, 1974; Ambraseys, 1995; Ambraseys and Adams, 1996; Ballore, 1917; Barrera, 1931; Böse, 1902, 1908; Calderón de la Barca, 1843; Camacho, 1991; Camacho, 1925; Camacho and Viquez, 1993; Case *et al.*, 1971; Córdoba *et al.*, 1994; Courboulex *et al.*, 1997; Crawford, 1898, 1902a, b; Cruz and Wyss, 1983; Dewey and Algermisson, 1974; Dewey and Saurez, 1991; Durham, 1931; Earthquake Engineering Research Institute, 1973; Figueroa, 1963a; Gonzalez-Ruiz and McNally, 1988; Goodyear, 1980; Hurriaga, 1997; Kanamori and Kikuchi, 1993; Kirkpatrick, 1920; Konig, 1952; Kovach, 1997; Larde y Larin, 1957; Leeds, 1986; Levin, 1940; Lonsdale and Klitgord, 1978; MacDonald and Johnston, 1919; Maldonado-Koerdell, 1958; Martin del Pozzo *et al.*, 1997; Mendoza and Nishenko, 1989; Miyamura, 1976, 1980; Montero, 1988; Montessus de Ballore, 1884, 1885; National Earthquake Information Centre, 1973; Nishenko and Singh, 1987a, b; Olsen, 1987; Orozco y Berra, 1887; Orozco y Berra, 1960; Ortiz *et al.*, 1998; Pacheco *et al.*, 1997; Peraldo and Montero, 1994; Perrey, 1854, 1857; Rockstroh, 1883; Rojas *et al.*, 1993; Rutten and van Raadshooven, 1940; Singh and Pacheco, 1994; Singh *et al.*, 1981, 1984, 1987, 1989, 1996; Sornette and Knopoff, 1997; Sultan, 1931; Sutch, 1979, 1981; Tristan, 1916; Troll, 1970; Urbina and Camacho, 1913; Viquez, 1994; Waitz and Urbina, 1919; Wolters, 1986.

**Caribbean loop**  Bernard and Lambert, 1988; Bonini *et al.*, 1984; Brown, 1907; Cornish, 1908; Cotilla, 1998; Deville, 1867; Dewey, 1972; Esquivel, 1995; Estorch, 1852; Fischer and McCann, 1984; Fuller, 1907; Hall, 1922; Herd *et al.*, 1981; Jordan, 1975; Lomnitz and Hashizume, 1985; Lorie, 1852; Mann, 1999; Mann and Burke, 1984; McCann, 1985; Molnar and Sykes, 1969; Page, 1930; Page, 1986; Poey y Aquirre, 1855, 1858; Reid and Taber, 1919, 1920; Robson, 1964; Rod, 1956; Scherer, 1912a, b; Sykes and Ewing, 1965; Taber, 1920a, 1922; Toiran, 2002; Tomblin and Robson, 1977; Wiggins-Grandison, 2001.

**South American and Andean zone**  Abe, 1979; Barazangi and Isacks, 1976; Barriga, 1951; Beck and Ruff, 1989; Beck and Nishenko, 1990; Branner, 1912; Campbell, 1914; Chatelain *et al.*, 1997; Collier and Sater, 1996; Dewey and Spence, 1979;

Dorbath *et al.*, 1990; Gerth, 1949; Giesecke and Silgado, 1981; Hasegawa and Sacks, 1981; Hayes, 1966; Humboldt and Bonpland, 1822; James, 1971; Kadinsky-Cade *et al.*, 1985; Kanamori and McNally, 1982; Kelleher, 1972; Langer and Spence, 1995; Lomnitz and Cabré, 1968; Lonsdale, 1978; Milne, 1880; Montessus de Ballore, 1911–16; Montessus de Ballore, 1912, 1917; Nishenko, 1985; Perrey, 1854, 1857; Polo, 1898, 1899, 1904; Ramirez, 1933, 1948, 1951, 1971, 1974; Rivera, 1983; Rudolph and Szirtes, 1911–12; Rudolph and Tams, 1907; Scholz, 1982, 1990, 1997; Sieberg, 1930; Silgado, 1968, 1973, 1992; Suárez *et al.*, 1983; Valencias, 1917; Willis, 1929.

***Continental and eastern seaboard*** Abbott, 1926; Aguilera, 1920; Anglin, 1984; Bent, 1992; Bollinger, 1977; Brasch, 1916; Broadhead, 1902; Clarke, 1969; Doig, 1990; DuBois and Smith, 1980; Dutton, 1889; Ebel, 1996; Gagnon, 1891; Goodfellow, 1888; Hodgson, 1925a, b; Jones, 1915; La Flamme, 1908.

## Chapter 3   Myths and legends

***Historical importance and role*** Andree, 1891; Bennett, 1991; Bode, 1977; Clark, 1953, 1955; Curtin, 1912; Cushing, 1896; Davis, 2000; Dubois, 1932; Eels, 1878; Frazer, 1930; Jacobs and Jacobs, 1959; Kroeber, 1976; Leet, 1948; McKenney and Hall, 1838–44; Minor and Grant, 1996; Mooney, 1896; Muser, 1978; Poole and Poole, 1962; Robertson, 1968; Snow, 1994; Swan, 1857; Tucker, 1956; Vitaliano, 1973; Viitanen, 1973; Williams, 1962; Williams, 1863.

***Aztecs, Maya, and Incas*** Alva Ixtlixóchitl, 1975; Amram, 1937, 1942; Anales de Tlatelolco, 1948; Baumann, 1963; Bierhorst, 1992a, b; Camus, 1987; Cline, 1944; Duran, 1971; Florescano, 1994; Freidel *et al.*, 1993; Goetz and Morley, 1950; Gubler and Bolles, 1997; Guiteras, 1961; Holland, 1963; Infante, 1986; Landa, 1938; Leon-Portilla, 1990; Lorenzo, 1979; Morris and von Hagen, 1993; Peñafiel, 1885; Recinos, 1950; Recinos *et al.*, 1953; Roys, 1967; Solis *et al.*, 1987; Spence, 1923; Sullivan, 1996; Tedlock, 1985; Thompson, 1970; Tozzer, 1907; Vaillant, 1962; Velázquez, 1975; Wisdom, 1940.

## Chapter 4   Earthquake effects

***Seismic intensities and ground accelerations*** Archuleta, 1982; Brune, 1996, 1999; Chávez and Castro, 1988; Clark, 1972; Davison, 1921; Dengler and McPherson, 1993; Evernden, 1975; Evernden *et al.*, 1981; Figueroa, 1963b; Freeman, 1932; Gasperini *et al.*, 1999; Gutenberg and Richter, 1942, 1956; Hanks and Johnson, 1976; Joyner and Boore, 1981; Neumann, 1954; Ohmachi and Midorikawa, 1992; Oldham, 1889; Richter, 1958a, b, 1959; Trifunac and Brady, 1975; Wald *et al.*, 1999; Wood and Neumann, 1931.

***Earthquake-resistant design*** Coburn and Spence, 1992; Freeman, 1932; Richter, 1958a; Wiegel, 1970.

***Directions of fall*** Akima, 1950; Anooshehpoor *et al.*, 1999; Benioff, 1938; Bolt, 1968; Bolt and Hansen, 1977; Clements, 1933; Ikegami and Kishinouye, 1947, 1949,

1950; Imamura, 1937; Kirkpatrick, 1927; Milne, 1885; Milne and Omori, 1893; Omori, 1900, 1903, 1907; Shi *et al.*, 1996; Takeo, 1998; Takeo and Ito, 1997; Weichert, 1994; Yamaguchi and Odaka, 1974.

*Maya architecture* Bernal, 1980; Brasseur de Bourbourg, 1866; Caso and Rubin de la Borbolla, 1936; Catherwood, 1844; Charnay, 1994; Coe, 1966; Kelly, 1982; Kidder *et al.*, 1946; Kramer, 1935; Lhuillier, 1952a, b, c, 1958a, b, c, d, 1962a, b; Lothrop, 1925; Maler 1908a, b, 1911; Marquina, 1928; Maudslay, 1889–1902; Merwin and Vaillant, 1932; Morley, 1917, 1920, 1932–1938, 1935, 1946; Morris *et al.*, 1931; Norman, 1843; Nuñez Chinchilla, 1963; Pollock, 1937; Potter, 1977; Proskouriakoff, 1946; Ramos, 1946; Roys, 1934; Ruppert, 1952; Sharer, 1994; Smith, 1940; Stephens, 1841, 1843; Thompson, 1904; Totten, 1926; Wauchope, 1938.

*Archaeological indicators of earthquake occurrences* Akyüz and Altunel, 2001; Altunel and Barka, 1996; Ellenblum *et al.*, 1998; Guidoboni, 1996, 2001; Karcz and Kafri, 1978, 1981; Marco *et al.*, 1997; Nur and Cline, 2000; Nur and Ron, 1997a, b; Rapp, 1986; Stewart and Buck, 2001; Stiros, 1996.

## Chapter 5    Earthquakes of Mexico

*Pre-conquest codices* Códices Mayas, 1976; Davoust, 1995; De Gruyter, 1946; Förstemann, 1880; Furst, 1978; Los Códices Mayas, 1985; Schele, 1982; Schellhas, 1904; Villacorta and Villacorta, 1933.

*Maya, Aztec, and Zapotec calendars* Barrera, 1980; Beyer, 1931; Caso, 1928; Chan, 1993; Infante, 1986; Leon-Portilla, 1990; Lorenzo, 1979; Marcus, 1992; Marcus and Flannery, 1996; Morley, 1975; Muser, 1978; Sidrys *et al.*, 1975; Weitzel, 1944; Wright, 1989.

*Documents of the colonial period* Anales de Tlatelolco, 1948; Berdan and Anawalt, 1997; Bierhorst, 1992a; Brotherston, 1995; Codex Mexicanus, 1934–40; Codex Vaticanus Nr. 3773 (Codex Vaticanus B), 1902; Códice de 1576, 1963; Códice Borgia, 1963; Códice Telleriano-Remensis, 1995; Cogulludo, 1957; Dibble, 1981; Edmonson, 1982; Goetz and Morley, 1950; Gubler and Bolles, 1997; Kingsborough, 1831–48; Kirchhoff *et al.*, 1976; Landa, 1938; Nuttall, 1903; Radin, 1920; Recinos, 1950; Recinos *et al.*, 1953; Roys, 1967; Tedlock, 1985; Tozzer, 1941; Velázquez, 1975.

*Mention of early earthquakes* Acosta and Reynoso, 1996; Anales de Tlatelolco, 1948; Bierhorst, 1992b; Códice de 1576, 1963; Codex Mexicanus, 1934–1940; Códice Telleriano-Remensis, 1995; Figueroa, 1963a; Orozco y Berra, 1887; Rabiela *et al.*, 1987; Singh *et al.*, 1980, 1983, 1985; Sugawara, 1987; Suter *et al.*, 1996; Torquemada, 1969.

## Chapter 6    Earthquakes in the Maya empire

*Mayan setting and history* Andrews, 1975; Brasseur de Bourbourg, 1866; Catherwood, 1844; Charnay, 1887, 1994; Coe, 1966, 1984, 1992; Cogulludo, 1957;

Ferguson and Royce, 1977, 1984; Gann, 1926, 1927; Humboldt, 1995; Hunter, 1974; Instituto Panamericano de Geografia e Historia (IPGH), 1939 Kelly, 1982; Marquina, 1928; Morley, 1946; Morley and Brainerd, 1956; Morley *et al.*, 1983; Moseley and Terry, 1980; Muser, 1978; Norman, 1843; Nuñez Chinchilla, 1963; Orozco y Berra, 1960; Pagden, 1975; Pasztory, 1997; Restall, 1998; Sabloff, 1990; Saville, 1918a; Sharer, 1974, 1994, 1996; Stephens, 1841, 1843; Thompson, 1954, 1970; Tozzer, 1907; Vogt, 1994; Weeks, 1993; Wolf, 1959.

*Archaeological studies* Adams *et al.*, 1981; Ashmore and Sharer, 1978; Bernal, 1980; Brunhouse, 1973, 1975; Carr and Hazard, 1961; Coe, 1965a, b; Diehl, 1983; Folan *et al.*, 1983; Galindo, 1835; Gordon, 1896, 1899; Graham, 1972, 1990; Gruning, 1930; Hammond, 1974; Hammond and Willey, 1979; Healy, 1980; Hewett, 1912; Howe, 1911; Joyce, 1914, 1929; Joyce *et al.*, 1927, 1928; Kidder *et al.*, 1946; Kramer, 1935; Le Count *et al.*, 2002; Linne, 1934; Longyear, 1944; Lothrop, 1924, 1925, 1933; Mackie, 1961, 1985; Maler, 1908a, b, 1911; Maudslay, 1883, 1886, 1889–1902; Maudslay and Maudslay, 1899; Merwin and Vaillant, 1932; Moctezuma, 1990; Morley, 1917, 1920, 1932–38, 1935; Morris, 1925, 1926, 1927, 1928; Morris and Morris, 1934; Morris *et al.*, 1931; Pollock, 1937; Potter, 1977; Ricketson and Ricketson, 1937; Roys, 1934; Ruppert, 1925, 1931, 1943, 1952; Ruppert *et al.*, 1954; Sansores, 1956; Satterthwaite, 1958, 1961; Saville, 1892; Scherzer, 1855; Seler, 1902–23, 1990; Sharer, 1969, 1978, 1990; Shook, 1952; Smith, 1929, 1940, 1950, 1972, 1982; Spinden, 1913; Stone, 1941; Stromsvík, 1941, 1947; Thompson, 1904; Thompson, 1939; Thompson *et al.*, 1932; Totten, 1926; Tozzer, 1957; Trik, 1939; Wauchope, 1938; Willey, 1973; Willey and Smith, 1969; Willey *et al.*, 1975.

*Earthquake occurrences* Almengor, 1994; Bevan and Sharer, 1983; Blom, 1935; Böse, 1902, 1903; Brune, 1968; Bucknam *et al.*, 1978; Chacón, 1909; Dewey and Saurez, 1991; Espinosa, 1976; Feldman, 1993; Güendel and Bunghum, 1995; Harlow *et al.*, 1993; Kanamori and Stewart, 1978; Langer and Bollinger, 1979; Leeds, 1986; Lomnitz and Schulz, 1996; Mackie, 1961, 1985; Morley, 1918; Plafker, 1976; Rockstroh, 1883; Rojas *et al.*, 1993; Sapper, 1902; Saville, 1892, 1918a; Seismological Notes; Taracena Flores, 1970; Trik, 1939; Wells, 1857; White, 1985; White and Harlow, 1993; Zilbermann de Lujan, 1987.

*Geologic and tectonic setting* Anderson *et al.*, 1973; Bateson and Hall, 1977; Bonini *et al.*, 1984; Bonis *et al.*, 1970; Burkart, 1978; Carr and Stoiber, 1977; Dengo and Bohnenberger, 1969; Dixon, 1956; Dixon *et al.*, 1998; Donnelly *et al.*, 1968; Erdlac and Anderson, 1982; Jordan, 1975; Kozuch, 1991; Kupfer and Godoy, 1967; Maldonado-Koerdell, 1958; Mann and Burke, 1984; Muehlberger and Ritchie, 1975; Nagle *et al.*, 1977; Schwartz *et al.*, 1979; Tournon and Alvarado, 1997; Vanek *et al.*, 2000; Vinson, 1962; Wadell, 1938; Ward *et al.*, 1985; Weidie, 1967.

*Volcanoes and eruptions* Anderson, 1908; Ballore, 1884, 1885; Bullard, 1976; Conyers, 1996; Crawford, 1902a, b; Dixon *et al.*, 1998; Feldman, 1993; Goodyear, 1880; Martin del Pozzo *et al.*, 1997; Martinez, 1978; McBirney and Williams, 1965; Merwin and Vaillant, 1932; Rockstroh, 1883; Sapper, 1925; Sheets, 1971a, 1971b, 1979, 1983, 1992; Simkin *et al.*, 1981; Villa Roiz, 1997; Viquez, 1994; Wood, 1987.

## Chapter 7    Earthquakes of Costa Rica, Panama, and Colombia

*General* Adamek *et al.*, 1988; Camacho, 1991; Camacho and Viquez, 1993; De Pinilla
and Toral, 1987; Jaggar, 1911a, b; Kirkpatrick, 1920, 1931; Mendoza and
Nishenko, 1989; Miyamura, 1980; Montero, 1988; Montero and Dewey, 1982;
Pennington, 1981; Peraldo and Montero, 1994; Plafker and Ward, 1992;
Schneider *et al.*, 1987; Spofford, 1911; Tryggvason and Lawson, 1970; Van der
Hilst and Mann, 1994; Viquez, 1994.

## Chapter 8    Earthquakes of Chile and Peru

*Cusco area earthquakes* Cabrera *et al.*, 1987; Cabrera and Sébrier, 1998; Eriksen
*et al.*, 1954; Esquivel y Navia, 1980; Wightman, 1990.
*Environmental and tectonic setting* Baumann, 1963; Bawden, 1996; Brüggen, 1943;
Farrington, 1983; Fiedel, 1987; Hayes, 1966; James, 1971; Keatinge, 1988;
MacNeish *et al.*, 1981; Mellafe, 1986; Morris and Von Hagen, 1993; Moseley, 1983,
1992; Moseley and Cordy-Collins, 1990; Moseley and Day, 1982; Myers, 1975;
Ortloff *et al.*, 1982, 1983; Peet, 1903; Rodriguez, 1990; Roosevelt, 1935;
Rostworowski de Diez Canseco, 1978–80; Shippee, 1932, 1933; Silverman, 1993;
Suárez *et al.*, 1983; Wilson, 1988.
*Landslides* Bode, 1989; Bolt *et al.*, 1975; Cluff, 1971; Keefer, 1983; Kojan and
Hutchinson, 1978; Lee and Duncan, 1975; Plafker *et al.*, 1971; Silgado, 1951;
Uhrhammer, 1996.
*Natural disasters and tectonic events* Amunátegui, 1882; Barazangi and Isacks,
1976; Beck and Ruff, 1989; Beck and Nishenko, 1990; Bird, 1987; Chatelain *et al.*,
1997; Dewey and Spence, 1979; Dorbath *et al.*, 1990; Duke and Leeds, 1963;
Espinosa *et al.*, 1977; Gajardo, 1970; Giesecke and Silgado, 1981; Husid *et al.*,
1977; Langer and Spence, 1995; Lomnitz, 1970b, 1971, 1983; Lomnitz and Cabré,
1968; Miano Pique, 1972; Montessus de Ballore, 1911–16, 1912, 1917; Nials *et al.*,
1979a, b; Nishenko, 1985; Oliver-Smith, 1986; Perrey, 1854, 1857; Polo, 1898,
1899, 1904; Rivera, 1983; St. Amand, 1961; Schwartz, 1988; Shimada *et al.*, 1991;
Shippee, 1932, 1933; Sieberg, 1930; Silgado, 1951, 1968, 1973, 1992; Thompson
*et al.*, 1985; Umlauff, 1915; Valencias, 1917; Weischet, 1963; Willis, 1929; Wright
and Mella, 1963; Wyss, 1978.
*Tsunami* Berninghausen, 1962; Bird, 1987; Darwin, 1839; Dillion and Oldale, 1978;
Heck, 1947; Lomnitz, 1970a; Plafker and Savage, 1970; Richards and Broeker,
1963; Shapiro *et al.*, 1998; Sievers *et al.*, 1963; Weigel, 1970.

## Chapter 9    Early California earthquakes

*Historical accounts* Agnew, 1978; Agnew and Sieh, 1978; Aldrich *et al.*, 1986;
Anderson and Bonin, 1987; Balderman *et al.*, 1978; Bancroft, 1886a, b, 1888;
Beanland and Clark, 1994; Bolton, 1926; Carpenter, 1921; Engelhardt, 1897,
1922; Freeman, 1932; Lawson *et al.*, 1908; Louderback, 1947; Oakeshott *et al.*,
1972; Smith and Teggart, 1910; Symposium, 1925; Toppozada and Borchardt,

1998; Tuttle and Sykes, 1992; Wood, 1955; Working Group on Californic Earthquake Probabilities (WGCEP), 1990.

*Paleoseismic investigations* Biasi and Weldon, 1994; Hall *et al.*, 1995; Hudnut and Sieh, 1989; Jacoby *et al.*, 1987, 1988, 1995; Lienkaemper and Prescott, 1989; Niemi, 1992; Niemi and Hall, 1992; Noller and Lightfoot, 1997; Noller *et al.*, 1994; Prentice and Schwartz, 1991; Sieh, 1977, 1978a, b, 1984; Sieh and Jahns, 1984; Sieh *et al.*, 1989; Smith, 1979; Yeats *et al.*, 1997; Youngs and Coppersmith, 1985.

## Chapter 10   Earthquakes of the North American Cordillera

*Basin and Range Earthquakes* Agnew, 1978; Balderman *et al.*, 1978; Beanland and Clark, 1994; Berry, 1916; Brumbaugh, 1980, 1984; Byerly, 1926; Callaghan and Gianella, 1935; Doser, 1985a, b; DuBois *et al.*, 1982; Eberhart-Phillips *et al.*, 1981; Eppley and Cloud, 1961; Gianella and Callaghan, 1934; Giardina, 1977; Hobbs, 1910; Jones, 1915; Page, 1935; Pearthree and Calvo, 1987; Pelton *et al.*, 1984; Reid, 1911; Richins *et al.*, 1987; Rogers *et al.*, 1983; Ryall, 1962; Slemmons *et al.*, 1965; Smith and Sbar, 1974; Smith, 1979; Staunton, 1918; Stein and Barrientos, 1985; Steinbrugge and Cloud, 1962; Stover, 1985; Tocher, 1956, 1962; Tocher *et al.*, 1957; Wallace, 1984; Wallace *et al.*, 1984; Williams and Tapper, 1953; Witkind *et al.*, 1962; Woollard, 1958; Yeats *et al.*, 1997.

*New Mexico and Texas earthquakes* Bilich *et al.*, 1998; Dumas *et al.*, 1980; Northrop, 1976.

*1887 Sonora, Mexico earthquake* Aguilera, 1901, 1920; Bull and Pearthree, 1988; Dubois and Smith, 1980; Gianella, 1960; Goodfellow, 1888; Hunt and Douglas, 1888; Lumholtz, 1912; MacDonald, 1918; Natali and Sbar, 1982; Russell, 1908; Sumner, 1977; Suter, 2001; Suter and Contreras, 2002.

*Death Valley seismic events* Bolt and Miller, 1975; Bulletins of the Berkeley Seismographic Stations, 1973–92; Given *et al.*, 1986, 1987, 1988, 1989; Gomberg, 1991, 1993; Harmsen, 1991, 1993a, b; Harmsen and Bufe, 1992; Harmsen and Rogers, 1986, 1987; Hileman *et al.*, 1973; Norris *et al.*, 1986a, b; Toppozada *et al.*, 1978; Wald *et al.*, 1990a, b; 1991, 1992, 1993, 1994, 1995, 1996, 1997.

*Death Valley general* Bennett *et al.*, 1997; Freund, 1974; Gan *et al.*, 2000; Hamilton, 1988; Hunt, 1975; Jennings *et al.*, 1962; Lingenfelder, 1986; Lingenfelder and Gash, 1984; Minster and Jordan, 1987; Nadeau, 1965; Noble and Wright, 1954; Thatcher *et al.*, 1999; Toppozada, 1975; Wernicke *et al.*, 1988a, b.

## Chapter 11   Earthquakes of eastern and Central America

*Eastern North America* Abbot, 1926; Anglin, 1984; Doig, 1990; Ebel, 1996; Ebel *et al.*, 1986; Frankel, 1995; Fujita and Sleep, 1991; Gagnon, 1891; Hodgson, 1925a, b, 1927, 1928, 1950; La Flamme, 1908; Lefebvre, 1928; Smith, 1962, 1966; Stevens, 1980, 1995; Street and Lacroix, 1979; Thwaites, 1896–1901; Winkler, 1992.

*1886 Charleston, South Carolina* Bollinger, 1972, 1973, 1974, 1977; Bollinger and
Visvanathan, 1977; Dutton, 1889; Freeman, 1932; Hanks and Johnston, 1992;
Louderback, 1944; Robinson and Talwani, 1983; Taber, 1914; Talwani, 1977,
1999; Talwani and Cox, 1985; Talwani and Schaeffer, 2001; Talwani and Sharma,
1999; Tarr *et al.*, 1981; Wood, 1945.

*New Madrid Earthquakes* Broadhead, 1902; Fuller, 1912; Grollimund and Zoback,
2001; Hermann *et al.*, 1978; Hough *et al.*, 2000; Johnston, 1996a, b; Johnston and
Nava, 1985; Johnston and Schweig, 1996; Kelson *et al.*, 1996; Li *et al.*, 1998; Lyell,
1849; Macelwane, 1930; McGee, 1892; Newman *et al.*, 1999; Nuttli, 1969, 1973,
1974, 1976, 1982, 1983; Nuttli and Zollweg, 1974; Obermeier, 1989; Obermeier
and Pond, 1999; Obermeier
*et al.*, 1990; O'Connell *et al.*, 1982; Penick, 1976; Reiter, 1990; Sampson, 1913;
Saucier, 1991; Schweig *et al.*, 1993; Shepard, 1905; Sibol *et al.*, 1987; Triep and
Sykes, 1997; Tuttle and Schweig, 1995, 1996; Tuttle *et al.*, 1996; Weber *et al.*,
1998; Wesnousky and Leffler, 1992; Zoback *et al.*, 1980.

*Texas Earthquakes* Byerly, 1934; Davis and Pennington, 1989; Davis *et al.*, 1989;
Doser, 1987; Dumas *et al.*, 1980; Ni *et al.*, 1981; Northrop and Sanford, 1972;
Yerkes and Castle, 1976.

## Chapter 12   Conclusion and speculations

*General* Acosta, 1968, 1973, 1976a, b; Figueroa, 1963a, b; Izumi and Sono, 1963;
Izumi and Terada, 1972; Jiménez *et al.*, 1999; Karcz and Kafri, 1978, 1981; King
*et al.*, 1994; Lhuillier, 1952a, b, c, 1958a, b, c, d, 1962a, b; Maudslay, 1889–1902;
Noller and Lightfoot, 1997; Ramos, 1946; Rapp, 1986; Seler, 1915; Sieh, 1996;
Singh *et al.*, 2000; Willey *et al.*, 1975.

# References

Abbott, C. D. 1926. "The St. Lawrence earthquake of February 28, 1925," *Bulletin of the Seismological Society of America*, **16**: 133–145.

Abe, K. 1979. "Size of great earthquakes of 1837–1974 inferred from tsunami data," *Journal of Geophysical Research*, **84**: 1561–1568.

Abe, K. and H. Kanamori. 1980. "Magnitudes of great shallow earthquakes from 1953 to 1977," *Tectonophysics*, **62**: 191–203.

Acosta, J. R. 1968. "Exploraciones en Palenque, 1967," *Departamento de Monumentos Prehispanicos, Informes, Instituto Nacional de Antropología e Historia*, **14**: 1–56.

1973. "Exploraciones y restauraciones en Palenque (1968–1970)," *Anales del Instituto Nacional de Antropología e Historia*, **3**(51): 21–70.

1976a. "Exploraciones en Palenque temporada (1973–1974)," *Anales del Instituto Nacional de Antropología e Historia*, **5**(53): 43–62.

1976b. "Exploraciones en Palenque durante 1972 (1974–1975)," *Anales del Instituto Nacional de Antropología e Historia*, **5**(53): 5–42.

Acosta, V. G. and G. S. Reynoso. 1996. *Los sismos en la historia de México*, Vol. 1. México: Fondo de Cultura Económica/Universidad Nacional Autónoma de México.

Adamek, S., C. Frohlich, and W. D. Pennington. 1988. "Seismicity of the Caribbean Nazca boundary: constraints on microplate tectonics of the Panama region," *Journal of Geophysical Research*, **93**: 2053–2075.

Adams, J. 1990. "Paleoseismicity of the Cascadia subduction zone: evidence from turbidites off the Oregon–Washington margin," *Tectonics*, **9**: 569–583.

Adams, R. E. W. 1973. "The collapse of Maya civilization, a review of previous theories," in *The Classic Maya Collapse*, ed. T. P. Culbert. Albuquerque: University of New Mexico Press, 21–34.

Adams, R. E. W. and W. D. Smith. 1977. "Apocalyptic visions: the Maya collapse and medieval Europe," *Archaeology*, **30**: 292–301.

Adams, R. E. W., W. E. Brown, Jr., and T. P. Culbert. 1981. "Radar mapping, archaeology and ancient Maya land use," *Science*, **213**: 1457–1463.

Agnew, D. C. 1978. "The 1852 Fort Yuma earthquake: two additional accounts," *Bulletin of the Seismological Society of America*, **68**: 1761–1762.

Agnew, D. C. and K. E. Sieh. 1978. "A documentary study of the felt effects of the great California earthquake of 1857," *Bulletin of the Seismological Society of America*, **68**: 1717–1729.

Aguilera, J. G. 1901. *Sobre las condiciones tectónicas de la República Mexicana.* Mexico: Oficina Tipografia de la Secretaria de Fomento.

　　1920. "The Sonora earthquake of 1887," *Bulletin of the Seismological Society of America*, **10**: 31–44.

Akima, T. 1950. "Experiments on the overturnings of circular columns by the aid of a shaking-table," *Bulletin of the Earthquake Research Institute (Tokyo)*, **28**: 333–348.

Akyüz, H. S. and E. Altunel. 2001. "Geological and archaeological evidence for post-Roman earthquake surface faulting at Cibyra, SW Turkey," *Geodinámica acta*, **14**: 95–101.

Aldrich, M. L., B. A. Bolt, A. E. Leviton, and P. U. Rodda. 1986. "The report of the 1868 Haywards earthquake," *Bulletin of the Seismological Society of America*, **76**: 71–76.

Algermissen, S. T., J. W. Dewey, C. J. Langer, and W. H. Dillinger. 1974. "The Managua earthquake of December 23, 1972: location, focal mechanism, and intensity distribution," *Bulletin of the Seismological Society of America*, **64**: 993–1004.

Allen, C. R. 1975. "Geologic criteria for evaluating seismicity," *Geological Society of America Bulletin*, **86**: 248–250.

Allen, C. R., P. St. Amand, C. F. Richter, and J. M. Nordquist. 1965. "Relationship between seismicity and geologic structure in the southern California region," *Bulletin of the Seismological Society of America*, **55**: 753–797.

Almengor, O. G. P. 1994. *La nueva Guatemala de la Asuncion y los terremotos de 1917–18.* Guatemala City: Universidad de San Carlos de Guatemala.

Altunel, E. and A. Barka. 1996. "Hierapolis'teki arkeosismik hasarin degerlendirilmesi" (evaluation of archaeoseismic damages at Hierapolis), *Geological Bulletin of Turkey*, **39**: 65–74.

Alva Ixtlixóchitl, F. de. 1975. *Obras históricas,* 2 vols, estudio introductorio y un apéndice documental por Edmundo O'Gorman, Serie de historiadores y cronistas de Indias 4: México: Universidad Nacional de Mexico.

Ambraseys, N. N. 1963. "The Buyin-Zara (Iran) earthquake of September, 1962," *Bulletin of the Seismological Society of America*, **53**: 705–740.

　　1995. "Magnitudes of Central American earthquakes 1898–1930," *Geophysical Journal International*, **121**: 545–556.

Ambraseys, N. N. and R. D. Adams. 1996. "Large Central American earthquakes 1898–1994," *Geophysical Journal International*, **127**: 665–692.

Ambraseys, N. N. and J. S. Tchalenko. 1969. "The Dasht-e Bayaz (Iran) earthquakes of August 31, 1968, a field report," *Bulletin of the Seismological Society of America*, **59**: 1751–1780.

Ambraseys, N. N. and A. Zátopek. 1968. "The Varto Üstükran (Anatolia) earthquake of 19 August 1966," *Bulletin of the Seismological Society of America*, **58**: 47–102.

　　1969. "The Mudurnu Valley, West Anatolia, Turkey earthquake of 22 July 1967," *Bulletin of the Seismological Society of America*, **59**: 521–577.

Amram, D. W., Jr. 1937. "Eastern Chiapas," *Geographical Review*, **27**: 19–36.

　　1942. "The Lacandon, last of the Maya," *El México antiguo*, **6**: 15–26.

Amunátegui, M. L. 1882. *El terremoto del 13 de mayo de 1647.* Santiago de Chile: Rafael Jover.

*Anales de Tlatelolco.* 1948. *Unos annales historicos de la nacion Mexicana y codice de Tlatelolco.* México: Antigua Libreria Robredo, de Jose Porrua e Hijos.

Anderson, H., S. Beanland, G. Blick, *et al.*, 1994. "The 1968 May 23 Inangahua, New Zealand earthquake: an integrated geological, geodetic, and seismological source model," *New Zealand Journal of Geology and Geophysics*, **37**: 59–86.

Anderson, J. G. and P. Bodin. 1987. "Earthquake recurrence models and historical seismicity in the Mexicali–Imperial Valley," *Bulletin of the Seismological Society of America*, **77**: 562–578.

Anderson, J. G., S. G. Wesnousky, and M. W. Stirling. 1996. "Earthquake size as a function of fault slip rate," *Bulletin of the Seismological Society of America*, **86**: 683–690.

Anderson, T. 1908. "The volcanoes of Guatemala," *Geographical Journal*, **31**: 473–489.

Anderson, T. H., B. Burkart, R. E. Clemons, O. Bohnenberger, and D. N. Blount. 1973. "Geology of the western Altos Cuchumatanes, northwestern Guatemala," *Geological Society of America Bulletin*, **84**: 805–826.

Andree, R. 1891. *Die Flutsagen, ethnographisch betrachtet.* Braunschweig: Friedrich Bieweg und Sohn.

Andrews, G. F. 1975. *Maya Cities, Placemaking and Urbanization.* Norman: University of Oklahoma Press.

Anglin, F. M. 1984. "Seismicity and faulting in the Charlevoix zone of the St. Lawrence valley," *Bulletin of the Seismological Society of America*, **74**: 595–603.

Anon. 2000. "Preliminary report on the 16 October 1999 M 7.1 Hector Mine, California earthquake," *Seismological Research Letters*, **71**: 11–23.

Anooshehpoor, A., T. Heaton, B. Shi, and J. Brune. 1999. "Estimates of the ground accelerations at Point Reyes Station during the 1906 San Francisco earthquake," *Bulletin of the Seismological Society of America*, **89**: 845–853.

Archuleta, R. 1982. "Analysis of near-source static and dynamic measurements from the 1979 Imperial Valley earthquake," *Bulletin of the Seismological Society of America*, **72**: 1927–1956.

Ashmore, W. and R. J. Sharer. 1978. "Excavations at Quiriguá, Guatemala: the ascent of an elite Maya center," *Archaeology*, **31**: 10–19.

Atwater, B. F. 1987. "Evidence for great Holocene earthquakes along the outer coast of Washington State," *Science*, **236**: 942–944.

  1992. "Geologic evidence for earthquakes during the past 2000 years along the Copalis River, southern coastal Washington," *Journal of Geophysical Research*, **97**: 1901–1919.

Atwater, B. F. and E. Hemphill-Haley. 1997. *Recurrence Intervals for Great Earthquakes of the Past 3500 Years at Northeastern Willapa Bay, Washington.* US Geological Survey Professional Paper 1576.

Atwater, B. F. and A. L. Moore. 1992. "A tsunami about 1000 years ago in Puget Sound, Washington," *Science*, **258**: 1614–1617.

Balderman, M. A., C. A. Johnson, D. G. Miller, and D. L. Schmidt. 1978. "The 1852
    Fort Yuma earthquake," *Bulletin of the Seismological Society of America*, **68**: 699–710.
Ballore, C. M. 1917. "The Mexican earthquake of November 12, 1912," *Bulletin of the
    Seismological Society of America*, **7**: 31–33.
Bancroft, H. H. 1886a. *History of California*, Vol. I, *1542–1800*. San Francisco: The History
    Company, Publishers.
    1886b. *The History of California*, Vol. II, *1801–1824*. San Francisco: The History
    Company, Publishers.
    1888. *The History of California*, Vol. VI, *1848–1859*. San Francisco: The History
    Company, Publishers.
Barazangi, M. and B. L. Isacks. 1976. "Spatial distribution of earthquakes and
    subduction of the Nazca plate beneath South America," *Geology*, **4**: 686–692.
Barrera, D. T. 1931. *El temblor del 14 de Enero de 1931*. México: Universidad Nacional de
    Mexico, Instituto de Geologia.
Barrera, V. A. 1980. *Diccionario maya*. Merida: Cordemex.
Barriga, V. M. 1951. *Los terremotos en Arequipa 1582–1868. Documentos de los Archivos de
    Sevilla y Arequipa*. Arequipa: Biblioteca Arequipa.
Bateson, J. H. and I. J. S. Hall. 1977. *The Geology of the Maya Mountains, Belize*. London:
    HM Stationery Office.
Baumann, H. 1963. *Gold and Gods of Peru*. New York: Pantheon Books.
Bawden, G. 1996. *The Moche*. Cambridge: Blackwell.
Beal, C. H. 1915. "The earthquake in the Imperial Valley, California," *Bulletin of the
    Seismological Society of America*, **5**: 130–149.
Beanland, S. and M. Clark. 1994. *The Owens Valley Fault Zone, Eastern California, and
    Surface Rupture Associated with the 1872 Earthquake*. US Geological Survey Bulletin
    1982.
Beck, S. L. and S. P. Nishenko. 1990. "Variations in the mode of great earthquake
    rupture along the Central Peru subduction zone," *Geophysical Research Letters*, **17**:
    1969–1972.
Beck, S. L. and L. J. Ruff. 1989. "Great earthquakes and subduction along the Peru
    Trench," *Physics of the Earth and Planetary Interiors*, **57**: 199–224.
Benioff, H. 1938. "The determination of the extent of faulting with application to
    the Long Beach earthquake," *Bulletin of the Seismological Society of America*, **28**:
    77–84.
Bennett, J. G. 1991. *Hazard*. Santa Fe: Bennett Books.
Bennett, R. A., B. P. Wernicke, J. L. Davis, *et al.*, 1997. "Global positioning constraints
    on fault slip rates in the Death Valley region, California and Nevada,"
    *Geophysical Research Letters*, **24**: 3073–3076.
Bent, A. L. 1992. "A re-examination of the 1925 Charlevoix, Quebec earthquake,"
    *Bulletin of the Seismological Society of America*, **82**: 2097–2113.
Berberian, M. 1979. "Earthquake faulting and bedding thrust associated with the
    Tabas-E-Golshan (Iran) earthquake of September 16, 1978," *Bulletin of the
    Seismological Society of America*, **69**: 1861–1887.

Berberian, M. and D. Papastamatiou. 1978. "Khurgu (North Bandar Abbas, Iran) earthquake of March 21, 1977: a preliminary field report and a seismotectonic discussion," *Bulletin of the Seismological Society of America*, **68**: 411–428.

Berdan, F. and P. Anawalt. 1997. *The Essential Codex Mendoza*. Berkeley: University of California Press.

Berg, J. W., Jr. and C. D. Baker. 1963. "Oregon earthquakes, 1841 through 1958," *Bulletin of the Seismological Society of America*, **53**: 95–108.

Bernal, I. 1980. *A History of Mexican Archaeology*. London: Thames and Hudson.

Bernard, P. and J. Lambert. 1988. "Subduction and seismic hazard in the northern Lesser Antilles: revision of the historical seismicity," *Bulletin of the Seismological Society of America*, **78**: 1965–1983.

Berninghausen, W. H. 1962. "Tsunamis reported from the west coast of South America 1562–1960," *Bulletin of the Seismological Society of America*, **52**: 915–922.

Berry, S. L. 1916. "An earthquake in Nevada," *Mining and Scientific Press*, **113**: 52–53.

Bevan, B. and R. J. Sharer. 1983. "Quiriguá and the earthquake of February 4, 1976," in *Quiriguá Reports, II*, University Monograph 49. Pennsylvania: University of Pennsylvania, 110–117.

Beyer, H. 1931. "The true zero date of the Maya," *Maya Research*, **3**: 202–204.

Biasi, G. P. and R. Weldon, II. 1994. "Quantitative refinement of calibrated $^{14}$C distributions," *Quaternary Research*, **41**: 1–18.

Bierhorst, J. 1992a. *Codex Chimalpopoca: the text in Nahuatl with a Glossary and Grammatical Notes*. Tucson: University of Arizona Press.

(trans.). 1992b. *History and Mythology of the Aztecs, the Codex Chimalpopoca*. Tucson: University of Arizona Press.

Bilich, A., S. Clark, B. Creighton, and C. Frolich. 1998. "Felt reports from the Alice, Texas earthquake of 24 March 1997," *Seismological Research Letters*, **69**: 117–122.

Bird, R. McK. 1987. "A postulated tsunami and its effects on cultural development in the Peruvian Early Horizon," *American Antiquity*, **52**: 285–303.

Blom, F. 1935. "The ruins of Copán and the earthquake," *Maya Research*, **2**: 291–292.

Bode, B. 1977. "Disaster, social structure, and myth in the Peruvian Andes: the genesis of an explanation," *Annals of the New York Academy of Sciences*, **293**: 246–274.

1989. *No Bells to Toll, Destruction and Creation in the Andes*. New York: Charles Scribner's Sons.

Bollinger, G. A. 1972. "Historical and recent seismic activity in South Carolina," *Bulletin of the Seismological Society of America*, **62**: 851–864.

1973. "Seismicity of the southeastern United States," *Bulletin of the Seismological Society of America*, **63**: 1785–1808.

1974. "Errata to Seismicity of the southeastern United States," *Bulletin of the Seismological Society of America*, **64**: 733.

1977. *Reinterpretation of the Intensity Data for the 1886 Charleston, South Carolina Earthquake*. US Geological Survey Professional Paper 1028-B, 17–32.

Bollinger, G. A. and T. R. Visvanathan. 1977. "The seismicity of South Carolina prior to 1886," in *Studies Related to the Charleston, South Carolina Earthquake of 1886: A Preliminary Report*, ed. D. W. Rankin. US Geological Survey Professional Paper 1028, 33–42.

Bolt, B. A. 1968. "The focus of the 1906 California earthquake," *Bulletin of the Seismological Society of America*, **58**: 457–472.

Bolt, B. A. and R. A. Hansen. 1977. "The upthrow of objects in earthquakes," *Bulletin of the Seismological Society of America*, **67**: 1415–1427.

Bolt, B. A. and R. D. Miller. 1975. *Catalogue of Earthquakes in Northern California and Adjoining Areas, 1 January 1910–31 December 1972*. Berkeley: Seismographic Stations, University of California.

Bolt, B. A., W. Horn, G. Macdonald, and R. Scott. 1975. *Geological Hazards: Earthquakes, Tsunamis, Volcanoes, Avalanches, Landslides, Floods*. Berlin: Springer-Verlag.

Bolton, H. E. (ed.). 1926. *Historical Memoirs of New California by Fray Francisco Palóu*, 4 vols. Berkeley: University of California Press.

Bonilla, M. G., R. G. Mark, and J. J. Lienkaemper. 1984. "Statistical relations among earthquake magnitude, surface rupture length, and surface fault displacement," *Bulletin of the Seismological Society of America*, **74**: 2379–2411.

Bonini, W. E., R. B. Hargraves, and R. Shagam (eds.). 1984. "The Caribbean–South American plate boundary and plate tectonics," *Geological Society of America Memoir* 162.

Bonis, S., O. H. Bohnenberger, and G. Dengo. 1970. *Mapa geologico de la república de Guatemala, escala 1:500 000*. Guatemala City: Instituto Geográfico Nacional.

Böse, E. 1902. "Sur les régions des tremblements de terre au Mexique," in *Memorias de la Sociedad Científica "Antonio Alzate*," Vol. XVIII. México: Imprenta del Gobierno en el Ex-Arzobispado, 159–184.

1903. "Los temblores de Zanatepec, Oaxaca a fines de septiembre de 1902. Estado actual de volcán tacaná, Chiapas," *Parergones del Instituto Geológico de México*, **1**(1): 5–25.

1908. "El temblor del 14 de abril de 1907," *Parergones del Insituto Geologico de Mexico*, **2**: 135–258.

Bradford, D. C. 1935. "Seismic history of the Puget-Sound Basin," *Bulletin of the Seismological Society of America*, **25**: 138–153.

Branner, J. C. 1912. "Earthquakes in Brazil," *Bulletin of the Seismological Society of America*, **2**: 105–117.

Brasch, F. E. 1916. "An earthquake in New England during the Colonial period (1755)," *Bulletin of the Seismological Society of America*, **6**: 26–42.

Brasseur de Bourbourg, C. E. 1866. *Palenque et autres ruines de l'ancienne civilisation du Mexique*. Paris: Bertrand.

Brewer, L. 1992. "Preliminary damage and intensity survey (Landers)," *Earthquakes and Volcanoes*, **23**: 219–226.

Broadhead, G. C. 1902. "The New Madrid earthquake," *American Geologist*, **30**: 76–87.

Brotherston, G. 1995. *Painted Books from Mexico*. London: British Museum Press.

Brown, C. W. 1907. "The Jamaican earthquake," *Popular Science Monthly*, **70**: 385–403.

Brown, R. D., Jr. P. L. Ward, and G. Plafker. 1973. *Geologic and Seismological Aspects of the Managua, Nicaragua, Earthquakes of December 23, 1972*. US Geological Survey Professional Paper 838.

Brüggen, J. 1943. *Contribución de la geología sísmica de Chile*. Santiago: Imprensa Universitaria.

Brumbaugh, D. S. 1980. "Analysis of the Williams, Arizona earthquake of November 4, 1971," *Bulletin of the Seismological Society of America*, **70**: 885–891.

1984. "A report of the Flagstaff, Arizona earthquake of 6 December 1981," *Bulletin of the Seismological Society of America*, **74**: 2041–2044.

Brune, J. N. 1968. "Seismic moment, seismicity and rate of slip along major fault zones," *Journal of Geophysical Research*, **73**: 777–784.

1996. "Precariously balanced rocks and ground-motion maps for southern California," *Bulletin of the Seismological Society of America*, **86**: 43–54.

1999. "Precarious rocks along the Mojave section of the San Andreas fault, California: constraints on ground motion from great earthquakes," *Seismological Research Letters*, **70**: 29–33.

2002. "Precarious-rock constraints on ground motion from historic and recent earthquakes in southern California," *Bulletin of the Seismological Society of America*, **92**: 2602–2611.

Brunhouse, R. L. 1973. *In Search of the Maya: the First Archaeologists*. Albuquerque: University of New Mexico Press.

1975. *Pursuit of the Ancient Maya: Some Archaeologists of Yesterday*. Albuquerque: University of New Mexico Press.

Bucknam, R. C., G. Plafker, and R. V. Sharp. 1978. "Fault movement (afterslip) following the Guatemala earthquake of February 4, 1976," *Geology*, **6**: 170–173.

Bull, W. B. and P. A. Pearthree. 1988. "Frequency and size of quaternary surface ruptures of the Pitaycachi fault, northeastern Sonora, Mexico," *Bulletin of the Seismological Society of America*, **78**: 956–978.

Bullard, F. M. 1976. *Volcanoes of the Earth*. Austin: University of Texas Press.

*Bulletins of the Seismographic Stations of the University of California, Berkeley*. 1973–92. Vols 43–62.

Burkart, B. 1978. "Offset across the Polochic fault of Guatemala and Chiapas, Mexico," *Geology*, **6**: 328–332.

Byerly, P. 1926. "The Montana earthquake of June 28, 1925, GMCT," *Bulletin of the Seismological Society of America*, **16**: 209–265.

1934. "The Texas earthquake of August 16, 1931," *Bulletin of the Seismological Society of America*, **24**: 81–99 and 303–325.

Cabrera, J. and M. Sébrier. 1998. "Surface rupture associated with a 5.3-mb earthquake: the 5 April 1986 Cuzco earthquake and kinematics of the Chincheros–Qorichocha faults of the High Andes, Peru," *Bulletin of the Seismological Society of America*, **88**: 242–255.

Cabrera, J., M. Sébrier, and J. L. Mercier. 1987. "Active normal faulting in high plateaus of Central Andes: the Cusco region (Peru)," *Annales tectonicae*, **2**: 116–138.

Calderón de la Barca, F. 1843. *Life in Mexico During a Residence of Two Years in that Country.* Boston: C. C. Little and J. Brown.

Callaghan, E. and V. P. Gianella. 1935. "The earthquake of January 30, 1934, at Excelsior Mountains, Nevada," *Bulletin of the Seismological Society of America*, **25**: 161–168.

Camacho, E. 1991. "The Puerto Armuelles earthquake (southwestern Panama) of July 18, 1934," *Revista geológica de América Central (Costa Rica)*, **13**: 1–13.

Camacho, E. and V. Viquez. 1993. "Historical seismicity of the North Panama Deformed Belt," *Revista geológica de América Central (Costa Rica)*, **15**: 49–64.

Camacho, H. 1925. "Apuntes acerca de la actividad actual del Popocatépetl en relación con la sismología," *Anales del Instituto Geológico de México*, **2**: 38–48.

Campbell, L. 1914. "Arequipa earthquakes registered during the year 1913," *Bulletin of the Seismological Society of America*, **4**: 81–87.

Camus, A. 1987. *American Journals.* New York: Paragon House Publishers.

Carpenter, F. A. 1921. "Early records of earthquakes in southern California," *Bulletin of the Seismological Society of America*, **11**: 1–3.

Carr, M. J. and R. L. Stoiber. 1977. "Geologic setting of some destructive earthquakes in Central America," *Geological Society of America Bulletin*, **88**: 151–156.

Carr, R. F. and J. E. Hazard. 1961. "Map of the ruins of Tikal, El Petén, Guatemala," *Tikal Reports, Museum Monographs* 11. Philadelphia: The University Museum.

Carydis, P. G. et al., 1982. *The Central Greece earthquakes of February–March 1981.* Washington, DC: National Academy Press.

Case, J. E., L. G. Duran, A. Lopez, and W. R. Moore. 1971. "Tectonic investigations in western Colombia and eastern Panama," *Geological Society of America Bulletin*, **82**: 2685–2712.

Caso, A. 1928. *Las estelas zapotecas.* Mexico: Talleres Graficos de la Nacion.

Caso, A. and D. F. Rubin de la Borbolla. 1936. *Exploraciones en Mitla 1934–1935*, Instituto Panamericano de Geografia e Historia, Publication 21.

Cassidy, J. F., R. M. Ellis, and G. C. Rogers. 1988. "The 1918 and 1957 Vancouver Island earthquakes," *Bulletin of the Seismological Society of America*, **78**: 617–635.

Catherwood, F. 1844. *Views of Ancient Monuments in Central America, Chiapas, and Yucatan.* New York: Barlett and Welford.

Chacón, B. 1909. "Los temblores del Año Pasado en Copán," letter to R. E. Durón, in *Revista de la Universidad, Tegucigalpa, Honduras*, **1**(6): 369–371.

Chan, R. P. 1993. *El lenguaje de las piedras.* Mexico: Fondo de Cultura Económica.

Charnay, D. 1887. *The Ancient Cities of the New World.* New York: Harper and Brothers. 1994. *Ciudades y ruinas americanas, reuindas y fotografiadas.* México: Banco de México.

Chatelain, J., B. Guillier, P. Guéguen, and F. Bondoux. 1997. "The $M_w$ 7.7 Nasca (Peru) earthquake, November 12, 1996: a repetition of the 1942 event?" *Seismological Research Letters*, **68**: 917–922.

Chávez, M. and R. Castro. 1988. "Attenuation of modified Mercalli intensity with distance in Mexico," *Bulletin of the Seismological Society of America*, **78**: 1875–1884.

Clark, E. E. 1953. *Indian Legends of the Pacific Northwest.* Berkeley: University of California Press.

1955. "George Gibbs' account of Indian mythology in Oregon and Washington territories," *Oregon Historical Quarterly,* **56**: 293–325.

Clark, M. M. 1972. *Intensity of Shaking Estimated from Displaced Stones.* US Geological Survey Professional Paper 787, 49–57.

Clarke, B. 1969. "America's greatest earthquake," *Reader's Digest,* **94**: 110–114.

Clements, T. 1933. "Notes on the fall of columns during the Long Beach earthquake," *Science,* **78**: 100–101.

Clements, T. and L. Clements. 1953. "Evidence of Pleistocene man in Death Valley," *Geological Society of America Bulletin,* **64**: 1189–1204.

Cline, H. 1944. "Lore and deities of the Lacandon Indians, Chiapas, Mexico," *Journal of American Folklore,* **57**: 107–115.

Cloud, W. K. 1967. "Intensity map and structural damage, Parkfield, California earthquake of June 27, 1966," *Bulletin of the Seismological Society of America,* **57**: 1161–1178.

Cluff, L. S. 1971. "Peru earthquake of May 31, 1970: engineering geology observations," *Bulletin of the Seismological Society of America,* **61**: 511–534.

Coburn, A. and R. Spence. 1992. *Earthquake Protection.* Chichester, England: Wiley.

*Codex Mexicanus.* 1934–1940. "Codex mexicanus: unos anales históricos de la nación mexicana," in E. Mengin (ed.), *Baessler Archiv,* **22**(2–3): 67–168 and **23**(4): 115–139.

*Codex Vaticanus Nr. 3773 (Codex Vaticanus B),* 1902. E. Seler (ed.). Berlin: Privately printed by Duke of Loubat.

*Códice de 1576 (Códice Aubin), Historia de la Nacion Mexicana.* 1963. Madrid: Ediciones Jose Porrua Turanzas.

*Códice Borgia.* 1963. México DF: Fondo de Cultura Económica.

*Códices Mayas.* 1976. A. Villacorta, C. and C. B. Villacorta (eds.). Guatemala City: Tipografia Nacional.

*Códice Telleriano-Remensis.* 1995. E. Q. Keber (ed.). Austin: University of Texas Press.

Coe, M. D. 1966. *The Maya.* New York: Praeger.

1984. *Mexico.* London: Thames and Hudson.

1992. *Breaking the Maya Code.* London: Thames and Hudson.

Coe, W. R. 1965a. "Tikal, Guatemala: an emergent Maya civilization," *Science,* **147**: 1401–1419.

1965b. "Tikal: ten years of study of a Maya ruin in the lowlands of Mexico," *Expedition,* **8**: 5–56.

Coffman, J. L. and C. A. von Hake (eds.). 1973a. "Earthquake history of the United States," *US Department of Commerce,* Publication 41-1.

1973b. *United States Earthquakes 1971.* US Department of Commerce.

1975. *United States Earthquakes 1973.* US Department of Commerce.

Cogolludo, D. L. de. 1957. *Historia de Yucatan* (1654). Mexico City: Editorial Academia Literaria.

Cole, S. C., B. F. Atwater, P. T. McCutcheon, J. K. Stein, and E. Hemphill-Haley. 1996. "Earthquake-induced burial of archaeological sites along the southern Washington coast about AD 1700," *Geoarchaeology*, **11**(2): 165–177.

Collier, S. and W. Sater. 1996. *A History of Chile, 1808–1994.* Cambridge: Cambridge University Press.

Conyers, L. B. 1996. "Archaeological evidence for dating the Loma Caldera eruption, Ceren, El Salvador," *Geoarchaeology*, **11**: 377–391.

Cooke, C. W. 1931. "Why the Mayan cities of the Petén district, Guatemala, were abandoned," *Journal of the Washington Academy of Sciences*, **21**: 283–287.

Corbett, E. J. and C. E. Johnson. 1982. "The Santa Barbara, California, earthquake of 13 August 1978," *Bulletin of the Seismological Society of America*, **72**: 2201–2226.

Córdoba, C., A. L. Martin del Pozzo, and J. Lopez Camacho. 1994. "Paleolandforms and volcanic impact on the environment of prehistoric Cuicuilco, southern Mexico City," *Journal of Archaeological Sciences*, **21**: 585–596.

Cornish, V. 1908. "The Jamaica earthquake," *Geographical Journal*, **31**: 245–276.

Cotilla, M. O. 1998. "An overview on the seismicity of Cuba," *Journal of Seismology*, **2**: 323–335.

Courboulex, F., M. A. Santoyo, J. F. Pacheco, and S. K. Singh. 1997. "The 14 September 1995 (M = 7.3) Copala, Mexico, earthquake: a source study using teleseismic, regional, and local data," *Bulletin of the Seismological Society of America*, **87**: 999–1010.

Cowgill, G. 1964. "The end of classic Maya culture: a review of recent evidence," *Southwestern Journal of Anthropology*, **20**: 145–159.

Craine, E. R. and R. C. Reindorp. 1979. *The Codex Perez and the Book of Chilam Balam of Maní.* Norman: University of Oklahoma Press.

Crawford, J. 1898. "Recent severe seismic disturbances in Nicaragua," *American Geologist*, **22**: 56–58.

　1902a. "List of the most important volcanic eruptions and earthquakes in western Nicaragua within historic time," *American Geologist*, **30**: 111–113.

　1902b. "Additions to the list of Nicaraguan volcanic eruptions in historic time," *American Geologist*, **30**: 395–396.

Cruz, G. and M. Wyss. 1983. "Large earthquakes, mean sea level, and tsunamis along the Pacific coast of Mexico and Central America," *Bulletin of the Seismological Society of America*, **73**: 553–570.

Culbert, T. P. (ed.). 1973. *The Classic Maya Collapse.* Albuquerque: University of New Mexico Press.

Curtin, J. 1912. *Myths of the Modocs.* Boston: Little, Brown, and Company.

Cushing, F. H. 1890. "Preliminary notes on the origin, working hypothesis and primary researches of the Hemenway southwestern archaeological expedition," *Congrès International des Americanistes compte-rendu de la septième session, Berlin, 1888*: 151–194.

　1896. "Outlines of Zuñi creation myths," in *13th Annual Report of the Bureau of American Ethnology for the Years 1891–1892*. Washington DC: US Government Printing Office, 321–347.

Darwin, C. 1839. *Narrative of the Surveying Voyages of His Majesty's Ships Adventure and Beagle, Between the Years 1826 and 1836, Describing their Examination of the Southern Shores of South America and the Beagle's Circumnavigation of the Globe*, Vol. 3. London: Henry Colburn.

Davis, N. Y. 2000. *The Zuñi Enigma*. New York: W. W. Norton and Company.

Davis, S. D. and W. D. Pennington. 1989. "Induced seismic deformation in the Cogdell oil field of west Texas," *Bulletin of the Seismological Society of America*, **79**: 1477–1494.

Davis, S. D., W. D. Pennington, and S. M. Carlson. 1989. *A Compendium of Earthquake Activity in Texas*. Geological Circular 89-3, Bureau of Economic Geology, University of Texas at Austin.

Davison, C. 1921. "On scales of seismic intensity and on the construction and use of isoseismal lines," *Bulletin of the Seismological Society of America*, **11**: 95–130.
    1936. *Great Earthquakes*. London: T. Murby.

Davoust, M. 1995. *L'écriture maya et son déchiffrement*. Paris: CNRS Editions.

De Gruyter, W. J. 1946. *A New Approach to Maya Hieroglyphs*. Amsterdam: Uitgeverij H. J. Paris.

Dengler, L. and R. McPherson. 1993. "The 17 August 1991 Honeydew earthquake, north coast California: a case for revising the Modified Mercalli Scale in sparsely populated areas," *Bulletin of the Seismological Society of America*, **83**: 1081–1094.

Dengo, G. and O. Bohnenberger. 1969. "Structural development of northern Central America," *American Association of Petroleum Geologists*, Memoir II: 203–220.

De Pinilla, V. V. and J. Toral. 1987. "Sismicidad histórica sentida en el istmo de Panamá," *Revista Geofisica*, **27**: 135–165.

Deville, C. S. 1867. "Sur le tremblement de terre du 18 November 1867 aux Antilles," *Comptes rendus de l'Académie des Sciences, Paris*, **65**: 1110–1114.

Dewey, J. W. 1972. "Seismicity and tectonics of western Venezuela," *Bulletin of the Seismological Society of America*, **62**: 1711–1751.
    1994. "Intensities and isoseismals (Northridge)," *Earthquakes and Volcanoes*, **25**: 85–93.

Dewey, J. and S. T. Algermisson. 1974. "Seismicity of the Middle America arc trench system near Managua, Nicaragua," *Bulletin of the Seismological Society of America*, **64**: 1033–1048.

Dewey, J. and G. Saurez. 1991. "Seismotectonics of Middle America," in *Neotectonics of North America*, 323–338. Boulder: Geological Society of America.

Dewey, J. W. and W. Spence. 1979. "Seismic gaps and source zones of recent large earthquakes in coastal Peru," *Pure and Applied Geophysics*, **117**: 1148–1171.

Dibble, C. E. 1981. *Codex en Cruz*. Salt Lake City: University of Utah Press.

Diehl, R. A. 1983. *Tula*. London: Thames and Hudson.

Dillion, W. P. and R. N. Oldale. 1978. "Late Quaternary sea level curve; reinterpretation based on glacio-tectonic influence," *Geology*, **6**: 56–60.

Dixon, C. G. 1956. *Geology of Southern British Honduras*. Belize: Printed by Authority of His Excellency the Governor.

Dixon, T. H., F. Farina, C. DeMets, P. Jansma, P. Mann, and E. Calais. 1998. "Relative motion between the Caribbean and North American plates and related boundary zone deformation from a decade of GPS observations," *Journal of Geophysical Research*, **103**: 15 157–15 182.

Doig, R. 1990. "2300 year history of seismicity from silting events in Lake Tadoussac, Charlevoix, Quebec," *Geology*, **18**: 820–823.

Donnelly, T. W., D. C. Crane, and B. Burkart. 1968. "Geologic history of the landward extension of the Bartlett Trough: some preliminary notes," *Transactions of the 4th Caribbean Geological Conference, Trinidad*: 225–228.

Dorbath, L., A. Cisternas, and C. Dorbath. 1990. "Assessment of the size of large and great historical earthquakes in Peru," *Bulletin of the Seismological Society of America*, **80**: 551–576.

Doser, D. I. 1985a. "The 1983 Borah Peak, Idaho and 1959 Hebgen Lake, Montana earthquakes: models for normal fault earthquakes in the intermountain seismic belt," *US Geological Survey Open File Report* 85-290: 368–384.

1985b. "Source parameters and faulting processes of the 1959 Hebgen Lake, Montana earthquake sequence," *Journal of Geophysical Research*, **90**: 4537–4555.

1987. "The August 16, 1931, Valentine, Texas earthquake: evidence for normal faulting in west Texas," *Bulletin of the Seismological Society of America*, **77**: 2005–2017.

Downes, G. L. 1995. *Atlas of Isoseismal Maps of New Zealand Earthquakes*. Lower Hutt; New Zealand: Institute of Geological and Nuclear Sciences Ltd.

Dubois, C. 1932. "Tolowa notes," *American Anthropologist*, **34**: 248–262.

DuBois, S. M. and A. W. Smith. 1980. *The 1887 Earthquake in San Bernardino Valley, Sonora: Historic Accounts and Intensity Patterns in Arizona*, Special Paper 3, State of Arizona Bureau of Geology and Mineral Technology.

DuBois, S. M., A. W. Smith, N. K. Nye, and T. A. Nowak, Jr. 1982. *Arizona Earthquakes, 1776–1980*, Bulletin 193, State of Arizona Bureau of Geology and Mineral Technology.

Duke, C. M. and D. J. Leeds. 1963. "Response of soils, foundations, and earth structures to the Chilean earthquakes of 1960," *Bulletin of the Seismological Society of America*, **53**: 309–357.

Dumas, D. B., H. J. Dorman and G. V. Latham. 1980. "A re-evaluation of the August 16, 1931 Texas earthquake," *Bulletin of the Seismological Society of America*, **70**: 1171–1180.

Durán, F. D. 1971. *Book of the Gods and Rites and the Ancient Calendar*. Norman: University of Oklahoma Press.

Durham, H. W. 1931. "Managua earthquake of 1931," *Engineering News Record*, April 23, 1931.

Dutton, C. E. 1889. "The Charleston earthquake of August 31, 1886," *US Geological Survey: Ninth Annual Report, 1887–88*, 203–528.

Earthquake Engineering Research Institute. 1973. *Managua, Nicaragua Earthquake of December 23, 1972*. Oakland: Earthquake Engineering Research Institute.

Ebel, J. E. 1996. "The seventeenth-century seismicity of northeastern North America," *Seismological Research Letters*, **67**: 51–68.

Ebel, J., P. Somerville, and J. D. McIver. 1986. "A study of the source parameters of some large earthquakes of northeastern North America," *Journal of Geophysical Research*, **91**: 8231–8247.

Eberhart-Phillips, D., R. M. Richardson, M. L. Sbar, and R. B. Hermann. 1981. "Analysis of the 4 February 1976 Chino Valley, Arizona earthquake," *Bulletin of the Seismological Society of America*, **71**: 787–801.

Edmonson, M. S. 1982. *The Ancient Future of the Itza: the Book of Chilam Balam of Tizimin, translated and annotated by M. S. Edmonson.* Austin: University of Texas Press.

Eels, M. 1878. "Traditions of the 'deluge' among the tribes of the north-west," *American Antiquarian*, **1**(2): 70–72.

Eisenberg, A., R. Husid, and J. E. Luco. 1972. "The July 8, 1971 Chilean earthquake," *Bulletin of the Seismological Society of America*, **62**: 423–430.

Ellenblum, R., S. Marco, A. Agnon, T. Rockwell, and A. Boas. 1998. "Crusader castle torn apart by earthquake at dawn, 20 May 1202," *Geology*, **26**: 303–306.

Engelhardt, Fr Z. 1897. *The Franciscans in California.* Harbor Springs, MI: Holy Childhood Indian School.

1922. *San Juan Capistrano Mission.* Los Angeles: The Standard Printing Co.

Eppley, R. A. and W. K. Cloud. 1961. *United States Earthquakes 1959.* US Department of Commerce.

Erdlac, R. J. Jr. and T. H. Anderson. 1982. "The Chixoy–Polochic fault and its associated fractures in western Guatemala," *Geological Society of America Bulletin*, **93**: 57–67.

Ericksen, G. E., J. F. Concha, and E. Silgado. 1954. "The Cusco, Peru earthquake of May 21, 1950," *Bulletin of the Seismological Society of America*, **44**: 97–112.

Espinosa, A. F. (ed.). 1976. *The Guatemala Earthquake of February 4, 1976, A Preliminary Report.* US Geological Survey Professional Paper 1002.

Espinosa, A. F., R. Husid, S. T. Algermissen, and J. de las Casas. 1977. "The Lima earthquake of October 3, 1974: intensity distribution," *Bulletin of the Seismological Society of America*, **67**: 1429–1439.

Esquivel, F. F. 1995. *Terremoto: los terremotos de Cartago en 1910.* Cartago: Urok Editores SA.

Esquivel y Navia, D. de 1980. *Noticias cronologicas de la gran ciudad del Cuzco* (2 Vols.) Lima: Fundacion Augusto N. Weise Banco Wiese LTDO.

Estorch, M. 1852. *Apuntes para la historia sobre el terremoto que tuvo lugar en Santiago de Cuba y otros puntos el 20 de agosto de 1852.* Santiago: Imprenta de D. Loreto Espinal.

Evernden, J. F. 1975. "Seismic intensities, 'size' of earthquakes, and related phenomena," *Bulletin of the Seismological Society of America*, **65**: 1287–1315.

Evernden, J. F., W. M. Kohler, and G. D. Clow. 1981. *Seismic Intensities of earthquakes of conterminous United States: Their Prediction and Interpretation.* US Geological Survey Professional Paper 1223.

Farrington, I. S. 1983. "The design and function of the intervalley canal: comments on a paper by Ortloff, Mosley and Feldman," *American Antiquity*, **48**: 360–375.

Feldman, L. H. 1993. *Mountains of Fire, Lands that Shake, Earthquakes and Volcanic Eruptions in the Historic Past of Central America (1505–1899).* Culver City, CA: Labyrinthos.

Ferguson, W. M. and J. Q. Royce. 1977. *Maya Ruins of Mexico in Color.* Norman: University of Oklahoma Press.

    1984. *Maya Ruins in Central America.* Albuquerque: University of New Mexico Press.

Fiedel, S. J. 1987. *Prehistory of the Americas.* Cambridge: Cambridge University Press.

Figueroa, A. J. 1963a. "Historia sísmica y estadística de temblores de la costa occidental de México," *Boletín bibliográfico de geofisica y oceanografia americanos*, **3**: 107–134.

    1963b. "Isosistas de macrosismos Mexicanos," *Ingeniera*, **33**: 45–67.

Fischer, K. M. and W. R. McCann. 1984. "Velocity modeling and earthquake location in the northeast Caribbean," *Bulletin of the Seismological Society of America*, **74**: 1249–1262.

Florescano, E. 1994. *Memory, Myth, and Time in Mexico from the Aztecs to Independence.* Austin: University of Texas Press.

Folan, W. J., E. R. Kintz, and L. A. Fletcher. 1983. *Coba: A Classic Maya Metropolis.* New York: Academic Press.

Förstemann, E. 1880. *Die Maya-Handschrift der königlichen öffentlichen Bibliothek zu Dresden.* Leipzig: Verlag der Naumann'schen Lichtdruckerei.

Frankel, A. 1995. "Mapping seismic hazard in the central and eastern United States," *Seismological Research Letters*, **66**: 8–21.

Frazer, J. B. 1930. *Myths of the Origin of Fire.* London: Macmillan Co.

Freeman, J. R. 1932. *Earthquake Damage and Earthquake Insurance.* New York: McGraw Hill Book Company.

Freidel, D., L. Schele, and W. Parker. 1993. *Maya Cosmos.* New York: William Morrow and Company, Inc.

Freund, R. 1974. "Kinematics of transform and transcurrent faults," *Tectonophysics*, **21**: 93–134.

Fujita, K. and N. H. Sleep. 1991. "A re-examination of the seismicity of Michigan," *Tectonophysics*, **186**: 75–106.

Fuller, M. 1907. "Notes on the Jamaica earthquake," *Journal of Geology*, **15**: 696–721.

Fuller, M. L. 1912. *The New Madrid earthquake.* US Geological Survey Bulletin 494.

Furst, J. L. 1978. *Codex Vindobonensis Mexicanus I: A Commentary*, Institute for Mesoamerican Studies, State University of New York, Publication 4.

Gagnon, A. 1891. "Le tremblement de terre de 1663 dans la Nouvelle France," *Proceedings of the Royal Society of Canada*, **9**: 41–52.

Gajardo, E. 1970. "Isoseismal map of the Peru earthquake of May 31, 1970," *Bulletin of the Seismological Society of America*, **60**: 2096–2097.

Galindo, J. 1835. "The ruins of Copán," *Archaeologia Americana. Transactions and Collections of the American Antiquarian Society*, **2**: 543–550.

Gan, W., J. L. Svarc, J. C. Savage, and W. H. Prescott. 2000. "Strain accumulation across the eastern California shear zone at latitude 36° 30′ N," *Journal of Geophysical Research*, **105**: 16229–16236.

Gann, T. 1926. *Ancient Cities and Modern Tribes: Exploration and Adventure in Mayaland*. London: Duckworth.

  1927. *Maya Cities: a Record of Exploration and Adventure in Middle America*. London: Duckworth.

Ganse, R. A. and J. B. Nelson. 1981. "Catalog of significant earthquakes 2000 BC–1979," *US Department of Commerce, Report* SE-27.

Gasperini, P., F. Bernardini, G. Valensise, and E. Boschi. 1999. "Defining seismogenic sources from historical felt reports," *Bulletin of the Seismological Society of America*, **89**: 94–110.

Geller, R. J. and H. Kanamori. 1977. "Magnitude of great shallow earthquakes from 1904 to 1952," *Bulletin of the Seismological Society of America*, **67**: 587–598.

Gerth, H. 1949. "Die geologischen Verhältnisse des vernichtenden Erdbebens in Ecuador im August, dieses Jahres (1949)," *Geologische Rundschau*, **37**: 83–85.

Gianella, V. P. 1960. "Faulting in northeastern Sonora, Mexico, in 1887," *Geological Society of America Bulletin*, **71**: 2061.

Gianella, V. P. and E. Callaghan. 1934. "The Cedar Mountain, Nevada, earthquake of December 20, 1932," *Bulletin of the Seismological Society of America*, **24**: 345–377.

Giardina, S. 1977. "A regional seismic evaluation of Flagstaff, Arizona," *Bulletin of the Association of Engineering Geologists*, **14**: 89–103.

Giesecke, A. and E. Silgado. 1981. *Terremotos en el Peru*. Lima: Ediciones Rikchay.

Given, D. D., R. Norris, L. M. Jones, *et al.*, 1986. "The southern California network bulletin January–June 1986," *US Geological Survey Open File Report* 86-598: 1–28.

Given, D. D., L. K. Hutton, and L. M. Jones. 1987. "The southern California network bulletin July–December 1986," *US Geological Survey Open File Report* 87-488: 1–40.

Given, D. D., L. A. Wald, L. M. Jones, and L. K. Hutton. 1988. "The southern California network bulletin January–June 1987," *US Geological Survey Open Report* 88-409: 1–34.

  1989. "The southern California network bulletin July–December 1987," *US Geological Survey Open Report* 89-323: 1–34.

Goetz, D. and S. G. Morley (trans.). 1950. *Popol Vuh, the Sacred Book of the Ancient Quiché Maya*. Norman: University of Oklahoma Press.

Gomberg, J. 1991. "Seismicity and detection/location threshold in the southern Great Basin seismic network," *Journal of Geophysical Research*, **96**: 16 401–16 414.

  1993. "Correction to 'Seismicity and shear strain in the southern Great Basin of Nevada and California' by Joan Gomberg," *Journal of Geophysical Research*, **98**: 4473–4476.

Gonzalez-Ruiz, J. R. and K. C. McNally. 1988. "Stress accumulation and release since 1882 in Ometepec, Guerrero, Mexico: implication for failure mechanisms and risk assessments of a seismic gap," *Journal of Geophysical Research*, **93**: 6297–6317.

Goodfellow, G. E. 1888. "The Sonora earthquake," *Science*, **11**: 162–166.

Goodyear, W. A. 1880. *Earthquake and Volcanic Phenomenon December 1879 and January 1880 in the Republic of Salvador, Central America.* Panama: Star and Herald.

Gordon, F. R. and J. D. Lewis. 1980. "The Meckering and Calingiri earthquakes October 1968 and March 1970," *Geological Survey of Western Australia Bulletin* 126.

Gordon, G. B. 1896. "Prehistoric ruins of Copán, Honduras. *Memoirs of the Peabody Museum of Archaeology and Ethnology, Harvard University,* **1**(1): 1–48.

1899. "The ruined city of Copán," *American Geographical Society Bulletin,* **31**: 39–50.

Graham, J. A. 1972. "The hieroglyphic inscriptions and monumental art of Altar de Sacrificios," *Papers of the Peabody Museum of Archaeology and Ethnology, Harvard University,* **64**(2): 1–85.

1990. "Excavations at Seibal, monumental sculpture and hieroglyphic inscriptions," *Memoirs of the Peabody Museum of Archaeology and Ethnology,* **17**(1): 1–79.

Greensfelder, R. W. 1972. "Intensities and magnitude of the great Owens Valley earthquake of 1872," *California Geology,* March 1972: 60.

Grollimund, B. and M. D. Zoback. 2001. "Did deglaciation trigger intraplate seismicity in the New Madrid seismic zone," *Geology,* **29**: 175–178.

Gruning, E. L. 1930. "Report on the British Museum expedition to British Honduras, 1930," *Journal of the Royal Anthropological Institute of Great Britain and Ireland,* **60**: 477–483.

Gubler, R. and D. Bolles. 1997. *The Book of Chilam Balam of Na.* Lancaster, CA: Labyrinthos.

Güendel, F. and H. Bungum. 1995. "Earthquakes and seismic hazards in Central America," *Seismological Research Letters,* **66**(5): 19–25.

Guidoboni, E. 1996. "Archaeology and historical seismology: the need for collaboration in the Mediterranean region," in *Archaeoseismology,* eds. S. Stiros and R. E. Jones, Fitch Laboratory Occasional Paper 7, Oxford: British School at Athens, 7–13.

2001. "A case study in archaeoseismology: the collapses of the temples at Selinunte (south-western Sicily)," *EOS, Transactions of the American Geophysical Union,* **82**(47): 32.

Guiteras, H. C. 1961. *Perils of the Soul. The World View of a Tzotil Indian.* New York: Glencoe and New York.

Gutenberg, B. and C. F. Richter. 1941. "Seismicity of the earth," *Geological Society of America Special Paper* 34.

1942. "Earthquake magnitude, intensity, energy and acceleration," *Bulletin of the Seismological Society of America,* **32**: 163–191.

1949. *Seismicity of the Earth and Associated Phenomena.* Princeton: Princeton University Press.

1956. "Earthquake magnitude, intensity, energy, and acceleration," *Bulletin of the Seismological Society of America,* **46**: 105–143.

Haghipour, A. and M. Amidi. 1980. "The November 14 to December 25, 1979 Ghaenet earthquakes of northeast Iran and their tectonic implications," *Bulletin of the Seismological Society of America,* **70**: 1751–1757.

Hall, M. 1922. *Earthquakes in Jamaica from 1688 to 1919.* Kingston: Government Printing Office.

Hall, T. H., R. H. Wright, and K. B. Clahan. 1995. "Paleoseismic investigations of the San Andreas fault on the San Francisco peninsula, California," *Geomatrix Consultants: Final Technical Report.*

Hamilton, W. B. 1988. "Detachment faulting in the Death Valley region, California and Nevada," in *Geologic and hydrologic investigations of a potential nuclear waste disposal site at Yucca mountain, southern Nevada.* US Geological Survey Bulletin 1790: 51–86.

Hammond, N. (ed.). 1974. *Mesoamerican Archaeology: New Approaches.* London: Duckworth.

  1982. *Ancient Maya Civilization.* New Brunswick: Rutgers University Press.

Hammond, N. 1998. "New lights on the ancient Maya," Lecture, Archaeological Institute of America, Stanford University.

Hammond, N. and G. R. Willey (eds.) 1979. *Maya Archaeology and Ethnohistory.* Austin: University of Texas Press.

Hanks, T. C. and D. A. Johnson. 1976. "Geophysical assessment of peak accelerations," *Bulletin of the Seismological Society of America,* **66**: 959–968.

Hanks, T. C. and A. C. Johnston. 1992. "Common features of the excitation and propagation of strong ground motion for North American earthquakes," *Bulletin of the Seismological Society of America,* **82**: 1–23.

Hanks, T. C. and H. Kanamori. 1979. "A moment magnitude scale," *Journal of Geophysical Research,* **84**: 2348–2350.

Hanks, T. C., J. A. Hileman, and W. Thatcher. 1975. "Seismic moments of the larger earthquakes of the southern California region," *Geological Society of America Bulletin,* **86**: 1131–1139.

Hansen, A. F. and V. M. Chavez. 1973. "Isoseismal maps of the Managua December 23, 1972 earthquake," *EERI Conference Proceedings on the Managua, Nicaragua Earthquake,* **1**: 104–115.

Harlow, D. H., R. A. White, M. J. Rymer, and S. Alvarez G. 1993. "The San Salvador earthquake of 10 October 1986 and its historical context," *Bulletin of the Seismological Society of America,* **63**: 1143–1154.

Harmsen, S. C. 1991. "Seismicity and focal mechanisms for the southern Great Basin of Nevada and California in 1990," *US Geological Survey Open File Report* 91-367: 1–103.

  1993a. "Seismicity and focal mechanisms for the southern Great Basin of Nevada and California," *US Geological Survey Open File Report* 92-340: 1–100.

  1993b. "Preliminary seismicity and focal mechanisms for the southern Great Basin of Nevada and California: January 1992 through September 1992," *US Geological Survey Open File Report* 93-369: 1–213.

Harmsen, S. C. and C. G. Bufe. 1992. "Seismicity and focal mechanisms for the southern Great Basin and California: 1987 through 1989," *US Geological Survey Open File Report* 91-572: 1–216.

Harmsen, S. C. and A. M. Rogers. 1986. "Inferences about the local stress field from focal mechanisms: applications to earthquakes in the southern Great Basin of Nevada," *Bulletin of the Seismological Society of America*, **76**: 1560–1572.

1987. "Earthquake location data for the southern Great Basin of Nevada and California: 1984 through 1986," *US Geological Survey Open File Report* 87-596: 1–92.

Hasegawa, A. and I. S. Sacks. 1981. "Subduction of the Nazca plate beneath Peru as determined from seismic observations," *Journal of Geophysical Research*, **86**: 4971–4980.

Hauksson, E. 1987. "Seismotectonics of the Newport–Inglewood fault zone in the Los Angeles basin," *Bulletin of the Seismological Society of America*, **77**: 539–561.

Hauksson, E. and S. Gross. 1991. "Source parameters of the 1933 Long Beach earthquake," *Bulletin of the Seismological Society of America*, **81**: 81–98.

Haury, E. W. 1945. "The excavation of Los Muertos and neighboring ruins in the Salt River Valley, Southern Arizona," *Papers of the Peabody Museum of American Archaeology and Ethnology, Harvard University*, **24**(1): 1–223.

Hayes, D. E. 1966. "A geophysical investigation of the Peru–Chile trench," *Marine Geology*, **4**: 309–351.

Healy, P. F. 1980. *Archaeology of the Rivas Region, Nicaragua*. Waterloo, Ontario: Wilfrid Laurier University Press.

Heaton, T. H. and S. H. Hartzell. 1987. "Earthquake hazards on the Cascadia subduction zone," *Science*, **236**: 162–168.

Heaton, T. H. and H. Kanamori. 1984. "Seismic potential associated with subduction in the northwestern United States," *Bulletin of the Seismological Society of America*, **74**: 933–941.

Heck, N. H. 1947. "List of seismic sea waves," *Bulletin of the Seismological Society of America*, **37**: 269–286.

Herd, D. G., T. L. Youd, H. Meyer, *et al.*, 1981. "The great Tumaco, Colombia earthquake of 12 December 1979," *Science*, **211**, 441–445.

Hermann, R. B., S. Cheng, and O. W. Nuttli. 1978. "Archeoseismology applied to the New Madrid earthquakes of 1811 to 1812," *Bulletin of the Seismological Society of America*, **68**: 1751–1759.

Hewett, E. L. 1912. "The excavations of Quiriguá, Guatemala by the School of American Archaeology," *Bulletin of the Archaeological Institute of America*, **3**: 163–171.

Hileman, J. A., C. R. Allen, and J. M. Nordquist. 1973. *Seismicity of the Southern California Region, 1 January 1932 to 31 December 1972*. Pasadena: Seismological Laboratory, California Institute of Technology.

Hobbs, W. H. 1907. *Earthquakes: An Introduction to Seismic Geology*. New York: D. Appleton and Company.

1910. "The earthquake of 1872 in the Owens Valley, California," *Gerlands Beiträge zur Geophysik*, **10**: 352–385.

Hodell, D. A., J. H. Curtis, and M. Brenner. 1995. "Possible role of climate in the collapse of classic Maya civilization," *Nature*, **375**: 391–394.

Hodgson, E. A. 1925a. "The rotation effects of the St. Lawrence earthquake of February 28, 1925," *Journal of the Royal Astronomical Society of Canada*, **19**(6): 169–178.

1925b. "The St. Lawrence Earthquake, February 28, 1925," *Bulletin of the Seismological Society of America*, **15**: 84–99.

1927. "The marine clays of eastern Canada and their relation to earthquake hazards," *Journal of the Royal Astronomical Society of Canada*, **21**(7): 257–264.

1928. "The probable epicenter of the Saint Lawrence earthquake of February 5, 1663," *Journal of the Royal Astronomical Society of Canada*, **22**(8): 325–334.

1950. "The Saint Lawrence earthquake, March 1, 1925," *Publications of the Dominion Observatory (Ottawa)*, **7**: 363–436.

Holden, E. S. 1892. *Earthquakes in California in 1890 and 1891*. US Geological Survey Bulletin 95, 1–29.

1898. "A catalogue of earthquakes on the Pacific coast 1769 to 1897," *Smithsonian Miscellaneous Collections* 1087: 1–253.

Holland, W. R. 1963. *Medicina Maya en los Altos de Chiapas*. México, DF: Instituto Nacional Indigenista.

Hough, S. T., J. G. Armbruster, L. Seeber, and J. F. Hough. 2000. "On the Modified Mercalli intensities and magnitudes of the 1811–1812 New Madrid earthquakes," *Journal of Geophysical Research*, **105**: 23839–23864.

Housner, G. W. 1963. "The behavior of inverted pendulum structures during earthquakes," *Bulletin of the Seismological Society of America*, **53**: 403–417.

Howe, G. P. 1911. "The ruins of Tuloom," *American Anthropology*, **13**: 539–550.

Hudnut, K. and K. Sieh. 1989. "Behavior of the Superstition Hills fault during the past 330 years," *Bulletin of the Seismological Society of America*, **79**: 304–329.

Humboldt, A. de and A. Bonpland. 1822. *Personal Narrative of Travels to the Equinoctial Regions of the New Continent During the Years 1799–1804*. London: Longman, Hurst, Rees, Orme and Brown.

Humboldt, A. von. 1995. *Vistas de las cordilleras y monumentos de los pueblos indigenas de America*. Mexico: Smurfit Carton y Papel de Mexico.

Hunt, A. P. 1960. Archaeology Death Valley salt pan, California. *Utah University Anthropological Papers*, **47**.

Hunt, C. B. 1975. *Death Valley: Geology, Ecology, Archaeology*. Berkeley: University of California Press.

Hunt, C. B. and D. R. Mabey. 1966. *Stratigraphy and Structure, Death Valley, California*. US Geological Survey Professional Paper 494-A.

Hunt, T. S. and J. Douglas. 1888. "The Sonora earthquake of May 3, 1887," *Transactions of the Seismological Society of Japan*, **12**: 29–31.

Hunter, C. B. 1974. *A Guide to Ancient Maya Ruins*. Norman: University of Oklahoma Press.

Hurriaga, J. N. 1997. *El Popocatépetl ayer y hoy*. México: Editorial Diana.

Husid, R., A. F. Espinosa, and J. de las Casas. 1977. "The Lima earthquake of October 3, 1974: damage distribution," *Bulletin of the Seismological Society of America*, **67**: 1441–1472.

Ikegami, R. and F. Kishinouye. 1947. "A study on the overturning of rectangular columns in the case of the Nankai earthquake on December 21, 1946," *Bulletin of the Earthquake Research Institute (Tokyo)*, **25**: 49–56.

1949. "The acceleration of earthquake motion deduced from overturning of the gravestones in the case of the Imaichi earthquake on December 26, 1949," *Bulletin of the Earthquake Research Institute (Tokyo)*, **25**: 121–127.

1950. "The acceleration of earthquake motion deduced from overturning of the gravestones in the case of the Imaichi earthquake," *Bulletin of the Earthquake Research Institute (Tokyo)*, **28**: 121–128.

Imamura, A. 1937. *Theoretical and Applied Seismology*. Tokyo: Maruzen Company.

Infante, F. D. 1986. *La Estela de los soles o calendario azteca*. México, DF: Panorama Editorial.

Instituto Panamericano de Geografía e Historia (IPGH). 1939. *Atlas arqueológico de la Republica Mexicana*. Instituto Panamericano de Geografía e Historia Publication **41**. México: Instituto Panamericano de Geografía e Historia.

Isacks, B., J. Oliver, and L. R. Sykes. 1968. "Seismology and the new global tectonics," *Journal of Geophysical Research*, **73**: 5855–5899.

Izumi, S. and T. Sono. 1963. *Andes 2: Excavations at Kotosh, Peru 1960*. Tokyo: Kadokawa Publishing Co.

Izumi, S. and K. Terada. 1972. *Excavations at Kotosh, Peru: A Report on the Third and Fourth Expeditions*. Tokyo: University of Tokyo Press.

Jacobs, E. D. and M. Jacobs. 1959. *Nehalem Tillamook Tales*. Eugene: University of Oregon Books.

Jacoby, G. C., P. R. Sheppard, and K. E. Sieh. 1987. "Was the 8 December 1812 California earthquake produced by the San Andreas fault? Evidence from trees near Wrightwood," *Seismological Research Letters*, **58**: 14.

1988. "Irregular recurrence of large earthquakes along the San Andreas fault: evidence from trees," *Science*, **241**: 196–199.

Jacoby, G., G. Carver, and W. Wagner. 1995. "Trees and herbs killed by an earthquake ~300 yr ago at Humboldt Bay, California," *Geology*, **23**: 77–80.

Jacoby, G. C., D. E. Bunker, and B. E. Benson. 1997. "Tree-ring evidence for an AD 1700 Cascadia earthquake in Washington and northern Oregon," *Geology*, **25**: 999–1002.

Jaggar, T. A., Jr. 1911a. "The Costa Rica volcanoes, and the earthquakes of April 13 and May 4, 1910," *Association of Engineering Societies*, **46**: 49–62.

1911b. "The earthquake in Costa Rica," *Science Conspectus*, **1**(2): 33–40.

James, D. E. 1971. "Plate tectonics model for the evolution of the central Andes," *Geological Society of America Bulletin* **82**: 3325–3346.

Jennings, C. W., J. L. Burnett, and B. W. Troxel. 1962. *Geologic Map of California, Trona Sheet, Scale 1:250 000*. California: Division of Mines and Geology.

Jiménez, J. I., J. I. Villarreal, M. R. Centeno, *et al.*, 1999. "Tehuacán, Mexico, earthquake of June 15, 1999," *Seismological Research Letters*, **70**: 698–704.

Johnston, A. C. 1996a. "Seismic moment assessment of earthquakes in stable continental regions: I. Instrumental seismicity," *Geophysical Journal International*, **124**: 381–414.

1996b. "Seismic moment assessment of earthquakes in stable continental regions: III. 1811–1812 New Madrid, 1886 Charleston and 1755 Lisbon," *Geophysical Journal International*, **126**: 314–344.

Johnston, A. C. and S. J. Nava. 1985. "Recurrence rates and probability estimates for the New Madrid seismic zone," *Journal of Geophysical Research*, **90**: 6737–6753.

Johnston, A. C. and E. S. Schweig. 1996. "The enigma of the New Madrid earthquakes of 1811–1812," *Annual Review of Earth and Planetary Sciences*, **24**: 339–384.

Jones, J. C. 1915. "The Pleasant Valley, Nevada earthquake of October 2, 1915," *Bulletin of the Seismological Society of America*, **5**: 190–205.

Jordan, T. H. 1975. "The present-day motions of the Caribbean Plate," *Journal of Geophysical Research*, **80**: 4433–4439.

Joyce, T. A. 1914. *Mexican Archaeology.* London: Philip Lee Warner.

1929. "Report on the British Museum expedition to British Honduras, 1929," *Journal of the Royal Anthropological Institute of Great Britain and Ireland*, **59**: 439–459.

Joyce, T. A., J. C. Clark, and J. E. Thompson. 1927. "Report on the British Museum expedition to British Honduras, 1927," *Journal of the Royal Anthropological Institute of Great Britain and Ireland*, **57**: 295–323.

Joyce, T. A., T. Gann, E. L. Gruning, and R. C. E. Long. 1928. "Report on the British Museum expedition to British Honduras, 1928," *Journal of the Royal Anthropological Institute of Great Britain and Ireland*, **58**: 323–350.

Joyner, W. B. and D. M. Boore. 1981. "Peak horizontal accelerations and velocity from strong ground-motion records from the 1979 Imperial Valley, CA earthquake," *Bulletin of the Seismological Society of America*, **71**: 2011–2038.

Kadinsky-Cade, K., R. Reilinger, and B. Isacks. 1985. "Surface deformation associated with the November 23, 1977, Caucete, Argentina earthquake sequence," *Journal of Geophysical Research*, **90**: 12 691–12 700.

Kanamori, H. 1977. "The energy release in great earthquakes," *Journal of Geophysical Research*, **82**: 2981–2987.

1978. "Quantification of great earthquakes," *Tectonophysics*, **49**: 207–212.

1983. "Magnitude scale and quantification of earthquakes," *Tectonophysics*, **93**: 185–199.

Kanamori, H. and K. Abe. 1979. "Reevaluation of the turn-of-the century seismicity peak," *Journal of Geophysical Research*, **84**: 6131–6139.

Kanamori, H. and D. L. Anderson. 1975. "Theoretical basis of some empirical relations in seismology," *Bulletin of the Seismological Society of America*, **65**: 1073–1095.

Kanamori, H. and M. Kikuchi. 1993. "The 1992 Nicaragua earthquake: a slow tsunami earthquake associated with subducted sediments," *Nature*, **361**: 714–716.

Kanamori, H. and K. C. McNally. 1982. "Variable rupture mode on the subduction zone along the Ecuador–Colombia coast," *Bulletin of the Seismological Society of America*, **72**: 1241–1253.

Kanamori, H. and G. S. Stewart. 1978. "Seismological aspects of the Guatemala earthquake of February 4, 1976," *Journal of Geophysical Research*, **83**: 3427–3434.

Karcz, I. and U. Kafri. 1978. "Evaluation of supposed archaeoseismic damage in Israel," *Journal of Archaeological Science*, **5**: 237–253.

1981. "Studies in archaeoseismicity of Israel: Hisham's Palace, Jericho," *Israel Journal of Earth Sciences*, **30**: 12–23.

Keatinge, R. W. (ed.). 1988. *Peruvian Prehistory*. Cambridge: Cambridge University Press.

Keefer, D. K. 1983. "Landslides caused by earthquakes," *Geological Society of America Bulletin*, **95**: 406–421.

Kelleher, J. A. 1972. "Rupture zones of large South American earthquakes and some predictions," *Journal of Geophysical Research*, **77**: 2087–2103.

Kelly, J. 1982. *The Complete Visitor's Guide to Mesoamerican Ruins*. Norman: University of Oklahoma Press.

Kelson, K. I., G. D. Simpson, R. B. VanArsdale, C. C. Haradan, and W. R. Lettis. 1996. "Multiple Late Holocene earthquakes along the Reelfoot fault, central New Madrid seismic zone," *Journal of Geophysical Research*, **101**: 6151–6170.

Kerr, R. A. 1995. "Faraway tsunami hints at a really big northwest quake," *Science*, **267**: 962.

Kidder, A. V., J. D. Jennings, and E. M. Shook. 1946. *Excavations at Kaminaljuyu, Guatemala*, Carnegie Institution of Washington Publication 561. Washington, DC.

King, G., G. Bailey, and D. Sturdy. 1994. "Active tectonics and human survival strategies," *Journal of Geophysical Research*, **99**: 20 063–20 078.

Kingsborough, E. K. 1831–48. *Antiquities of Mexico*, 9 vols. London: James Moynes and Colnaghi, Son, and Company.

Kirchhoff, P., L. O. Güemes, and L. R. Garcia. 1976. *Historia Tolteca–Chichimeca*. México, DF: Instituto Nacional de Antropologia e Historia.

Kirkpatrick, P. 1927. "Seismic measurements by the overthrow of columns," *Bulletin of the Seismological Society of America*, **17**: 95–109.

Kirkpatrick, R. Z. 1920. "Earthquakes in Panama up to January 1, 1920," *Bulletin of the Seismological Society of America*, **10**: 121–128.

1931. "Earthquakes in Panama," *The Military Engineer*, **23**: 544–545.

Kojan, E. and J. N. Hutchinson. 1978. "Mayunmarca rockslide and debris flow, Peru," in *Rockslides and Avalanches*, Vol. 1. ed. B. Voight. Amsterdam: Elsevier, 315–361.

Konig, L. P. G. 1952. "Earthquakes in relation to their geographical distribution, Part 5, Central America and the Caribbean region," *Nederlandse Akademie van Wetenschappen, Proceedings, Series B*, **55**: 272–292.

Kovach, R. L. 1997. "Significant ancient Central American earthquakes," *EOS, Transactions of the American Geophysical Union*, **78**(46): 447.

1998. "Seismotectonics of Death Valley," *1998 Abstracts with Programs, Cordilleran Section, Geological Society of America*, **30**(5): 49.

Kozuch, M. J. (comp.) 1991. *Mapa geológico de Honduras, escala 1:500 000*. Tegucigalpa: Instituto Geográfico Nacional.

Kramer, G. T. 1935. "Roof-combs in the Maya area, an introductory study," *Maya Research*, **2**: 106–118.

Kroeber, A. L. 1976. *Yurok Myths*. Berkeley: University of California Press.

Kupfer, D. H. and J. Godoy. 1967. "Strike-slip faulting in Guatemala," *Transactions of the American Geophysical Union*, **48**: 215.

La Flamme, J. C. K. 1908. "Les tremblements de terre de la région de Québec," *Transactions of the Royal Society of Canada*, Third Series, 1907–1908, **1**: 157–183.

Landa, D. de. 1938. *Relación de las cosas de Yucatán*. Mérida: Edición Yucateca.

Lander, J. F. and W. K. Cloud. 1963. *United States earthquakes 1961*. US Department of Commerce.

Langer, C. J. and G. A. Bollinger. 1979. "Secondary faulting near the terminus of a seismogenic strike-slip fault: aftershocks of the 1976 Guatemala earthquake," *Bulletin of the Seismological Society of America*, **69**: 427–444.

Langer, C. J. and W. Spence. 1995. "The 1974 Peru earthquake series," *Bulletin of the Seismological Society of America*, **85**: 665–687.

Larde y Larin, J. 1957. *El Salvador*. San Salvador: Ministeria de Cultura Departmento Editorial.

Lawson, A. C., G. K. Gilbert, H. F. Reid, *et al.*, 1908. *The California Earthquake of April 18, 1906, Report of the State Earthquake Investigation Commission*, 2 vols and Atlas. Washington DC: Carnegie Institution of Washington.

Le Count, L., J. Yaeger, R. M. Leventhal, and W. Ashmore. 2002. "Dating the rise and fall of Xunantunich, Belize," *Ancient Mesoamerica*, **13**: 41–63.

Lee, K. L. and J. M. Duncan. 1975. *Landslide of April 25, 1974 on the Mantaro River, Peru*. Report of the National Academy of Sciences of the USA.

Leeds, D. J. 1986. "Catalog of Nicaraguan earthquakes," *Bulletin of the Seismological Society of America*, **64**: 1135–1158.

Leet, L. D. 1948. *Causes of Catastrophes*. New York: McGraw-Hill.

Lefebvre, J. H. 1928. "A vanished Niagara," *Bulletin of the Seismological Society of America*, **18**: 104–109.

Leon-Portilla, M. 1990. *Time and Reality in the Thought of the Maya*. Norman: University of Oklahoma Press.

Levin, S. B. 1940. "The Salvador earthquakes of December, 1936," *Bulletin of the Seismological Society of America*, **30**: 1–45.

Lhuillier, A. R. 1952a. "Exploraciones arqueológicas en Palenque: 1949," *Anales del Instituto Nacional de Antropologia e Historia*, **4**(32): 49–60.

  1952b. "Exploraciones arqueológicas en Palenque: 1950," *Anales del Instituto Nacional de Antropologia e Historia*, **5**(33): 25–46.

  1952c. "Exploraciones arqueológicas en Palenque: 1951," *Anales del Instituto Nacional de Antropologia e Historia*, **5**(33): 47–66.

  1958a. "Exploraciones arqueológicas en Palenque: 1953," *Anales del Instituto Nacional de Antropologia e Historia*, **10**(39): 69–116.

1958b. "Exploraciones arqueológicas en Palenque: 1954," *Anales del Instituto Nacional de Antropologia e Historia*, **10**(39): 117–184.

1958c. "Exploraciones arqueológicas en Palenque: 1955," *Anales del Instituto Nacional de Antropologia e Historia*, **10**(39): 185–240.

1958d. "Exploraciones arqueológicas en Palenque: 1956," *Anales del Instituto Nacional de Antropologia e Historia*, **10**(39): 241–299.

1962a. "Exploraciones arqueológicas en Palenque: 1957," *Anales del Instituto Nacional de Antropologia e Historia*, **14**: 35–90.

1962b. "Exploraciones arqueológicas en Palenque: 1958," *Anales del Instituto Nacional de Antropologia e Historia*, **14**: 91–112.

Li, Y., E. S. Schweig, M. P. Tuttle, and M. A. Ellis. 1998. "Evidence for large prehistoric earthquakes in the northern New Madrid seismic zone," *Seismological Research Letters*: **69**: 270–276.

Lienkaemper, J. J. and W. H. Prescott. 1989. "Historic surface slip along the San Andreas fault near Parkfield," *Journal of Geophysical Research*, **94**: 17 647–17 670.

Lingenfelder, R. E. 1986. *Death Valley and the Amargosa: A Land of Illusion*. Berkeley: University of California Press.

Lingenfelder, R. E. and K. R. Gash. 1984. *The Newspapers of Nevada, A History and Bibliography, 1854–1979*. Reno: University of Nevada Press.

Linne, S. 1934. *Researches at Teotihuacán, Mexico*. Stockholm: V. Pettersons.

Lomnitz, C. 1970a. "Major earthquakes and tsunamis in Chile during the period 1535 to 1955," *Geologische Rundschau*, **59**(3): 938–960.

1970b. "The Peru earthquake of May 31, 1970," *Bulletin of the Seismological Society of America*, **60**: 1413–1416.

1971. "Peru earthquake of May 31, 1970: some preliminary seismological results," *Bulletin of the Seismological Society of America*, **61**: 535–542.

1983. "On the epicenter of the great Santiago earthquake of 1647," *Bulletin of the Seismological Society of America*, **73**: 885–886.

Lomnitz, C. and R. Cabré. 1968. "The Peru earthquake of October 17, 1966," *Bulletin of the Seismological Society of America*, **58**: 645–661.

Lomnitz, C. and M. Hashizume. 1985. "The Popayán Colombia earthquake of 31 March 1983," *Bulletin of the Seismological Society of America*, **75**: 1315–1326.

Lomnitz, C. and R. Schulz. 1996. "The San Salvador earthquake of May 3, 1965," *Bulletin of the Seismological Society of America*, **56**: 561–575.

Longyear, J. M., III. 1944. "Archaeological investigations in El Salvador," *Memoirs of the Peabody Museum of Archaeology and Ethnology, Harvard University*, **9**(2): 1–115.

Lonsdale, P. 1978. "Ecuadorian subduction system," *American Association of Petroleum Geologists Bulletin*, **62**: 2454–2477.

Lonsdale, P. and D. Klitgord. 1978. "Structural and tectonic history of the eastern Panama Basin," *Geological Society of America Bulletin*, **89**: 1–9.

Lorenzo, A. 1979. *Uso e interpretacion del calendario Azteca*. México, DF: Miguel Angel Porrúa, SA.

Lorie, A. M. 1852. *El viernes de agosto en Cuba*. Santiago, Cuba: Imprenta de D. Miguel Antonio Martinez.

*Los códices mayas.* 1985. Chiapas, Mexico: Universidad Autonoma de Chiapas.

Lothrop, S. K. 1924. "Tulum, an archaeological study of the east coast of Yucatan," *Carnegie Institution of Washington,* Publication 335. Washington, DC.

1925. "The Architecture of the ancient Maya," *Architectural Record,* **57**: 491–509.

1933. "Atitlan: an archaeological study of the ancient remains on the borders of Lake Atitlan, Guatemala," *Carnegie Institution of Washington,* Publication 444. Washington, DC.

Louderback, G. D. 1944. "The personal record of Ada M. Trotter of certain aftershocks of the Charleston earthquake of 1886," *Bulletin of the Seismological Society of America,* **34**: 199–206.

1947. "Central California earthquakes of the 1830s," *Bulletin of the Seismological Society of America,* **37**: 33–74.

Lowe, J. W. G. 1985. *The Dynamics of Apocalypse.* Albuquerque: University of New Mexico Press.

Lumholtz, C. 1912. *New Trails in Mexico.* New York: Charles Scribner's Sons.

Lyell, C. 1849. *A Second Visit to the United States of North America,* 2 vols. New York: Harper and Brothers, Publishers.

Macdonald, B. 1918. "Remarks on the Sonora earthquake," *Bulletin of the Seismological Society of America,* **8**: 74–78.

Macdonald, D. F. and W. C. Johnston. 1919. "Isthmian earthquakes," *Canal Record,* **7**: 144–149.

Macelwane, J. B. 1930. "The Mississippi Valley earthquake problem," *Bulletin of the Seismological Society of America,* **20**: 95–98.

Mackie, E. W. 1961. "New light on the end of classic Maya culture at Benque Viejo, British Honduras," *American Antiquity,* **27**: 216–224.

1985. *Excavations at Xunantunich and Pomona, Belize, in 1959–60.* BAR International Series 251. Oxford.

MacNeish, R. S., A. G. Cook, L. G. Lumbreras, R. K. Vierra, and A. Nelken-Terner. 1981. *Prehistory of the Ayacucho Basin, Peru,* Vol. 2, *Excavations and Chronology.* Ann Arbor: The University of Michigan Press.

Maldonado-Koerdell, M. 1958. *Bibliografia geologica y paleontologica de America Central.* Instituto Panamerico de Geografia e Historia, Publication 204. México: Instituto Panamerico de Geographia e Historia.

Maler, T. 1908a. "Explorations of the upper Usumatsinta and adjacent regions," *Memoirs of the Peabody Museum of Archaeology and Ethnology, Harvard University,* **4**(1): 1–49.

1908b. "Explorations in the Department of Petén, Guatemala and adjacent regions: Topoxte, Yaxha, Benque Viejo, Naranjo," *Memoirs of the Peabody Museum of Archaeology and Ethnology, Harvard University,* **4**(2): 55–127.

1911. "Explorations in the Department of Petén, Guatemala: Tikal," *Memoirs of the Peabody Museum of Archaeology and Ethnology, Harvard University,* **5**(1): 3–91.

Mallet, R. and J. W. Mallet. 1858. "The Earthquake Catalogue of the British Association with the discussion, curves and maps, etc. (in 4 parts)," *Transactions of the British Association for the Advancement of Science, 1852–1858.*

Mann, P. 1999. "Caribbean sedimentary basins: classification and tectonic setting from Jurassic to present," in *Caribbean Basins. Sedimentary Basins of the World*, Vol. 4, ed. P. Mann, Amsterdam: Elsevier.

Mann, P. and K. Burke. 1984. "Neotectonics of the Caribbean," *Reviews of Geophysics and Space Physics*, **22**: 309–362.

Marco, S., A. Agnon, R. Ellenblum, *et al.*, 1997. "817-year-old walls offset sinistrally 2.1 m by the Dead Sea Transform Fault, Israel," *Journal of Geodynamics*, **24**: 11–20.

Marcus, J. 1992. *Mesoamerican Writing Systems.* Princeton: Princeton University Press.

Marcus, J. and K. V. Flannery. 1996. *Zapotec Civilization.* London: Thames and Hudson.

Marquina, I. 1928. *Estudio arquitectonico comparativo de los monumentos arqueologicos de Mexico.* México: Talleres Graficos de la Nacion.

Martin del Pozzo, A. L., C. Córdoba and J. López. 1997. "Volcanic impact on the southern basin of Mexico during the Holocene," *Quaternary International*, **43/44**: 181–190.

Martinez, H. M. A. 1978. *Cronologia sismica y eruptiva de republica de El Salvador a partir de 1820.* San Salvador: Ministerio de Obras Publicas, Centro de Investigaciones Geotechnicas.

Maudslay, A. C. and A. P. Maudslay. 1899. *A glimpse at Guatemala.* London: John Murray.

Maudslay, A. P. 1883. "Explorations in Guatemala, and examination of the newly-discovered Indian ruins of Quiriguá, Tikal, and the Usumacinta," *Proceedings of the Royal Geographical Society*, **5**: 185–203.

1886. "Exploration of the ruins and site of Copán, Central America," *Proceedings of the Royal Geographical Society*, **8**: 568–595.

1889–1902. *Archaeology. Biologia Centrali-Americana*, Vols 1–6. London: Porter.

McAdie, A. G. 1907. "Catalog of earthquakes on the Pacific coast 1897–1906," *Smithsonian Miscellaneous Collections*, 1721: 1–64.

McBirney, A. R. and H. Williams. 1965. *Volcanic History of Nicaragua.* Berkeley: University of California Press.

McCann, W. R. 1985. "On the earthquake hazards of Puerto Rico and the Virgin Islands," *Bulletin of the Seismological Society of America*, **75**: 251–262.

McGee, W. J. 1892. "A fossil earthquake," *Geological Society of America Bulletin*, **4**: 411–414.

McKenney, T. L. and J. Hall. 1838–1844. *History of the Indian Tribes of America*, 3 vols. Philadelphia: F. W. Greenough.

Mellafe, R. 1986. *Historia social de Chile y America.* Santiago: Editorial Universitaria.

Mendoza, C. and S. Nishenko. 1989. "The North Panama earthquake of 7 September 1882: evidence for active underthrusting," *Bulletin of the Seismological Society of America*, **79**: 1264–1269.

Merwin, R. E. and G. C. Vaillant. 1932. "The ruins of Holmul, Guatemala," *Memoirs of the Peabody Museum of Archaeology and Ethnology, Harvard University*, **3**(2): 1–143.

Miano Pique, C. 1972. *iiBasta!! La bomba Atómica Francesca, la contaminacíon atmosférica y los terremotos.* Lima: Tagrat.

Milne, J. 1880. "The Peruvian earthquake of May 9, 1877," *Transactions of the Seismological Society of Japan*, **2**: 50–96.

1885. "Seismic experiments," *Transactions of the Seismological Society of Japan*, **8**: 1–82.

1912. *A Catalogue of Destructive Earthquakes.* London: British Association for the Advancement of Science.

Milne, J. and F. Omori. 1893. "On the overturning and fracturing of brick and other columns by horizontally applied motion," *Seismological Journal of Japan*, **17**: 59–86.

Minor, R. and W. C. Grant. 1996. "Earthquake-induced subsidence and burial of Late Holocene archaeological sites, northern Oregon coast," *American Antiquity*, **61**(4): 772–781.

Minster, J. B. and T. H. Jordan. 1987. "Vector constraints on western US deformation from space geodesy, neotectonics, and plate motions," *Journal of Geophysical Research*, **92**: 4798–4804.

Miyamura, S. 1976. "Provisional magnitudes of middle American earthquakes not listed in the magnitude catalog of Gutenberg–Richter," *Bulletin of the International Institute of Seismology and Earthquake Engineering*, **14**: 41–46.

1980. *Sismicidad de Costa Rica.* San José: Editorial de Costa Rica.

Moctezuma, E. M. 1990. *Teotihuacán.* New York: Rizzoli.

Molnar, P. and L. R. Sykes. 1969. "Tectonics of the Caribbean and Middle America regions from focal mechanisms and seismicity," *Geological Society of America Bulletin*, **80**: 1639–1684.

Montandon, F. 1962. "Les megaseismes en Amerique: revue pour l'étude des calamités," *Bulletin de l'Union Internationale de Secours*, **38**: 57–97.

Montero, W. 1988. "Sismicidad histórica de Costa Rica," *Geofisica International*, **28**: 531–559.

Montero, W. and J. Dewey. 1982. "Shallow-focus seismicity, composite focal mechanism and tectonics of the Valle Central of Costa Rica," *Bulletin of the Seismological Society of America*, **72**: 1611–1626.

Montessus de Ballore, F. 1884. *Temblores y erupciones volcánicas en Centro-América.* San Salvador: Imprenta del Doctor Francisco Sagrini.

1885. "Sur les tremblements de terre et les éruptions volcaniques dans l'Amérique Centrale," *Comptes rendus de l'Académie des Sciences, Paris*, **100**: 1312–1315.

1911. *La sismologie moderne.* Paris: Armand Colin.

1911–16. *Historia sísmica de los Andes meridionales,* 6 vols. Santiago de Chile: Imprenta Cervantes.

1912. "Fenomeni luminosi speciale che avrebbero accompagnato il terremoto di Valparaíso del 16 di Agosto 1906," *Bolletino della Societá Sismologica italiana*, **16**: 1–28.

1917. "Terremoto del año 1582 en Arequipa," *Revista chilena de historia y geografia*, **24**: 3–39.

Mooney, J. 1896. "The ghost-dance religion and the Sioux outbreak of 1890," *Fourteenth Annual Report of the Bureau of American Ethnology for 1892-93*, **14**(2): 653–1136.

Morley, S. G. 1917. "The ruins of Tuloom, Yucatan," *American Museum Journal*, **17**: 190–204.

1918. "The Guatemala earthquake," *American Museum Journal*, **18**: 200–210.

1920. *The inscriptions at Copán*, Carnegie Institution of Washington Publication 219. Washington, DC.

1935. *Guidebook to the Ruins of Quiriguá*, Carnegie Institution of Washington, Supplementary Publication 16. Washington, DC.

1937–38. *The inscriptions of Petén*, 5 vols., Carnegie Institution of Washington Publication 437. Washington, DC.

1946. *The Ancient Maya*. Stanford: Stanford University Press.

1975. *An Introduction to the Study of the Maya Hieroglyphics*. New York: Dover Publications, Inc.

Morley, S. G. and G. W. Brainerd. 1956. *The Ancient Maya*, 3rd edition. Stanford: Stanford University Press.

Morley, S. G., G. W. Brainerd and R. J. Sharer. 1983. *The Ancient Maya.* Stanford: Stanford University Press.

Morris, C. and A. Von Hagen. 1993. *The Inka Empire*. New York: Abbeville Press.

Morris, E. H. 1925. "Report on the Temple of the Warriors," *Carnegie Institution of Washington Year Book* 24: 252–259.

1926. "Report on excavation of the Temple of the Warriors and the Northwest Colonnade," *Carnegie Institution of Washington Year Book* 25: 282–286.

1927. "Report on the Temple of the Warriors and the Northwest Colonnade," *Carnegie Institution of Washington Year Book* 26: 240–246.

1928. "Report on the excavation and repair of the Temple of the Warriors," *Carnegie Institution of Washington Year Book* 27: 293–297.

Morris, E. H. and A. A. Morris. 1934. "Quiriguá," *Carnegie Institution of Washington Year Book* 33: 86–89.

Morris, E. H., J. Charlot, and A. A. Morris. 1931. *The Temple of the Warriors at Chichen Itzá, Yucatan*, Carnegie Institution of Washington Publication 406. Washington, DC.

Moseley, E. H. and E. D. Terry (eds.). 1980. *Yucatan: A World Apart*. Mobile, AL: University of Alabama Press.

Moseley, M. E. 1983. "The good old days were better: agrarian collapse and tectonics," *American Anthropologist*, **85**: 773–799.

1992. *The Incas and their Ancestors.* London: Thames and Hudson.

Moseley, M. E. and A. Cordy-Collins (eds.). 1990. *The Northern Dynasties Kingship and State Craft in Chimor.* Washington, DC: Dumbarton Oaks Research Library and Collection.

Moseley, M. E. and K. C. Day (eds.). 1982. *Chan-Chan: Andean Desert City.* Albuquerque: University of New Mexico Press.

Muehlberger, W. R. and A. W. Ritchie. 1975. "Caribbean–Americas plate boundary in Guatemala and southern Mexico as seen on Skylab IV orbital photography," *Geology*, **3**: 232–235.

Murphy, L. M. and W. K. Cloud. 1954. *United States earthquakes 1952.* US Department of Commerce Coast and Geodetic Survey 773.

  1955. *United States earthquakes 1953.* US Department of Commerce Coast and Geodetic Survey 785.

  1956. *United States earthquakes 1954.* US Department of Commerce Coast and Geodetic Survey, 793.

  1957. *United States earthquakes 1955.* US Department of Commerce Coast and Geodetic Survey.

Murphy, L. M. and F. P. Ulrich. 1952. *United States earthquakes 1950.* US Department of Commerce Coast and Geodetic Survey 755.

Muser, C. 1978. *Facts and Artifacts of Ancient Middle America.* New York: E. P. Dutton.

Myers, J. S. 1975. "Vertical crustal movements of the Andes in Peru," *Nature,* **254**: 672–674.

Nadeau, R. 1965. *Ghost Towns and Mining Camps of California.* Los Angeles: Ward Ritchie Press.

Nagle, F., J. J. Stipp, and J. Rosenfeld. 1977. *Guatemala; Where Plates Collide: A Reconnaissance Guide to Guatemalan Geology.* Miami: Miami Geological Society.

Nason, R. 1982. *Seismic intensity studies in the Imperial Valley.* US Geological Survey Professional Paper 1254: 259–264.

Natali, S. G. and M. L. Sbar. 1982. "Seismicity in the epicentral region of the 1887 northeastern Sonoran earthquake, Mexico," *Bulletin of the Seismological Society of America,* **72**: 181–196.

National Earthquake Information Center. 1973. "Christmas in Managua," *Earthquake Information Bulletin,* **5**: 4–11.

Nelson, E. W. 1891. "The Panamint and Saline Valley Indians," *American Anthropologist,* **4**: 371–372.

Neumann, F. 1954. *Earthquake Intensity and Related Ground Motion.* Seattle: University of Washington Press.

Newman, A. V., S. Stein, J. Weber, *et al.,* 1999. "Slow deformation and low seismic hazard at the New Madrid seismic zone," *Science,* **284**: 619–621.

Ni, J. F., R. E. Reilinger, and L. D. Brown. 1981. "Vertical crustal movements in the vicinity of the 1931 Valentine, Texas, earthquake," *Bulletin of the Seismological Society of America,* **71**: 857–863.

Nials, F. L., E. E. Deeds, M. E. Moseley, *et al.,* 1979a. "El Niño: the catastrophic flooding of coastal Peru, pt. 1," *Field Museum of Natural History Bulletin,* **50**(7): 4–14.

  1979b. "El Niño: the catastrophic flooding of coastal Peru, pt. 2," *Field Museum of Natural History Bulletin,* **50**(8): 4–10.

Niemi, T. M. 1992. "Late Holocene slip rate, prehistoric earthquakes, and Quaternary neotectonics of the northern San Andreas fault, Marin County, California," PhD Dissertation, Stanford University.

Niemi, T. M. and N. T. Hall. 1992. "Late-Holocene slip rate and recurrence of great earthquakes on the San Andreas fault in northern California," *Geology,* **20**: 195–198.

Nishenko, S. P. 1985. "Seismic potential for large and great interplate earthquakes along the Chilean and southern Peruvian margins of South America: a quantitative reappraisal," *Journal of Geophysical Research*, **90**: 3589–3615.

Nishenko, S. P. and S. K. Singh. 1987a. "Relocation of the great Mexican earthquake of 14 January 1903," *Bulletin of the Seismological Society of America*, **77**: 256–259.

   1987b. "The Acapulco–Ometepec, Mexico earthquakes of 1907–1982: evidence for a variable recurrence history," *Bulletin of the Seismological Society of America*, **77**: 1359–1367.

Noble, L. F. and L. A. Wright. 1954. Geology of the central and southern Death Valley Region, California, *California Division of Mines and Geology Bulletin* 170, 143–160.

Noller, J. S. and K. G. Lightfoot. 1997. "An archaeoseismic approach and method for the study of active strike-slip faults," *Geoarchaeology*, **12**: 117–135.

Noller, J. S., W. R. Lettis, and G. D. Simpson. 1994. "Seismic archaeology: using human prehistory to date paleoearthquakes and assess deformation rates of active fault zones," in C. S. Prentice, D. P. Schwartz, and R. S. Yeats (eds.). *Proceedings of the Workshop on Paleoseismology. US Geological Survey Open File Report* 94-568: 138–140.

Norman, B. M. 1843. *Rambles in Yucatán.* New York: J. & H. G. Langley.

Norris, R. D., C. E. Johnson, L. M. Jones, and L. K. Hutton. 1986a. "The Southern California Network Bulletin January–June 1985," *US Geological Survey Open File Report* 86-96: 1–43.

Norris, R. D., L. M. Jones, and L. K. Hutton. 1986b. "The southern California network bulletin, July–December 1985," *US Geological Survey Open File Report* 86-337: 1–34.

Northrop, S. A. 1976. "New Mexico's earthquake history, 1849–1975," *Tectonics and Mineral Resources of Southwestern North America*, Special Publication 6, New Mexico Geological Society: 77–87.

Northrop, S. A. and A. R. Sanford. 1972. "Earthquakes of northeastern New Mexico and the Texas Panhandle," *New Mexico Geological Society, Annual Field Trip Conference Guidebook* 23: 148–160.

Nuñez Chinchilla, J. 1963. *Copán Ruins.* Tegucigalpa: Banco Central de Honduras.

Nur, A. and E. H. Cline. 2000. "Poseidon's horses: plate tectonics and earthquake storms in the Late Bronze Age in the Aegean and eastern Mediterranean," *Journal of Archaeological Science*, **27**: 43–63.

Nur, A. and H. Ron. 1997a. "Armageddon's earthquakes," *International Geology Review*, **39**: 532–541.

   1997b. "Earthquake! Inspiration for Armageddon," *Biblical Archaeology Review*, **23**: 49–55.

Nuttall, Z. 1903. *The Book of Life of the Ancient Mexicans. Part 1: Introduction and Facsimile.* Berkeley: University of California.

Nuttli, O. W. 1969. "Ground motion and magnitude relations for central United States earthquakes," *Earthquake Notes*, **40**: 14.

   1973. "The Mississippi valley earthquakes of 1811 and 1812: intensities, ground motion and magnitudes," *Bulletin of the Seismological Society of America*, **63**: 227–248.

1974. "Magnitude–recurrence relation for central Mississippi valley earthquakes," *Bulletin of the Seismological Society of America*, **64**: 1189–1207.

1976. "Comments on 'Seismic Intensities, 'Size' of Earthquakes and Related Parameters' by Jack F. Evernden," *Bulletin of the Seismological Society of America*, **66**: 331–338.

1982. *Damaging Earthquakes of the Central Mississippi Valley*. US Geological Survey Professional Paper 1236-B, 15–20.

1983. "Average seismic source–parameter relations for mid-plate earthquakes," *Bulletin of the Seismological Society of America*, **73**: 519–536.

Nuttli, O. W. and J. E. Zollweg. 1974. "The relation between felt area and magnitude for central United States earthquakes," *Bulletin of the Seismological Society of America*, **64**: 73–85.

Oakeshott, G. B., R. W. Greensfelder, and J. E. Kahle. 1972. "1872–1972 . . . one hundred years later," *California Geology*, **25**: 55–61.

Obermeier, S. F. 1989. *The New Madrid Earthquakes: An Engineering–Geologic Interpretation of Relict Liquefaction Features*. US Geological Survey Professional Paper 1336-B, 1–114.

Obermeier, S. F. and E. C. Pond. 1999. "Issues in using liquefaction features for paleoseismic analysis," *Seismological Research Letters*, **70**: 34–58.

Obermeier, S. F., R. B. Jacobson, J. P. Smoot, *et al.*, 1990. *Earthquake-induced Liquefaction Features in the Coastal Setting of South Carolina and in the Fluvial Setting of the New Madrid Seismic Zone*. US Geological Survey Professional Paper 1504, 1–44.

O'Connell, D. R., C. G. Bufe, and M. D. Zoback. 1982. *Microearthquakes and faulting in the area of New Madrid, Missouri-Reelfoot Lake, Tennessee*. US Geological Survey Professional Paper 1236-D, 31–38.

Officers of the Geological Survey of India and S. C. Roy. 1939. "The Bihar-Nepal earthquake of 1934," *Memoirs of the Geological Survey of India*, **73**: 1–391.

Ohmachi, T. and S. Midorikawa. 1992. "Ground-motion intensity inferred from upthrow of boulders during the 1984 western Nagano Prefecture, Japan, earthquake," *Bulletin of the Seismological Society of America*, **82**: 44–60.

Oldham, R. D. 1899. "Report on the great earthquake of 12th June 1897," *Memoirs of the Geological Survey of India*, **29**: 1–379.

Oliver-Smith, A. 1986. *The Martyred City*. Albuquerque: University of New Mexico Press.

Olsen, R. A. 1987. "The San Salvador earthquake of October 10, 1986: overview and context," *Earthquake Spectra*, **3**: 415–418.

Omori, F. 1900. "Seismic experiments on the fracturing and overturning of columns," *Publications of the Earthquake Investigation Committee in Foreign Languages*, **4**: 69–141.

1903. "On the overturning and sliding of columns," *Publications of the Earthquake Investigation Committee in Foreign Languages*, **12**: 8–28.

1907. "Preliminary note on the cause of the San Francisco earthquake of April 18, 1906," *Bulletin of the Imperial Earthquake Investigation Committee*, **1**: 7–25.

Orozco y Berra, J. 1887. "Efemérides séismicas Mexicanas," in *Memorias de la Sociedad Cientifica "Antonio Alzate,"* Vol. I, 303–541. Mexico: Imprenta del Gobierno en el Ex-Arzobispado.

Orozco y Berra, M. 1960. *Historia Antigua y de la Conquista de Mexico.* México, DF: Editorial Porrúa, SA.

Ortiz, M., S. K. Singh, J. Pacheco, and V. Kostoglodov. 1998. "Rupture length of the October 9, 1995 Colima-Jalisco earthquake ($M_w$8) estimated from tsunami data," *Geophysical Research Letters*, **25**: 2857–2860.

Ortloff, C. R., M. E. Moseley, and R. A. Feldman. 1982. "Hydraulic engineering aspects of the Chimu–Chicama–Moche intervalley canal," *American Antiquity*, **47**: 572–595.

  1983. "The Chicama–Moche intervalley canal: social explanations and physical paradigms," *American Antiquity*, **48**: 375–389.

Pacheco, J. F. and L. R. Sykes. 1992. "Seismic moment catalog of large shallow earthquakes, 1900–1989," *Bulletin of the Seismological Society of America*, **82**: 1306–1349.

Pacheco, J., S. K. Singh, J. Domínguez, *et al.*, 1997. "The October 9, 1995, Colima-Jalisco, Mexico earthquake ($M_w$8): an aftershock study and a comparison of this earthquake with those of 1932," *Geophysical Research Letters*, **24**: 2223–2226.

Pagden, A. R. 1975. *The Maya: Diego Landa's Account of the Affairs of Yucatán.* Chicago: J. Philip O'Hara, Inc.

Page, B. M. 1935. "Basin range faulting of 1915 in Pleasant Valley, Nevada," *Journal of Geology*, **43**: 690–707.

Page, S. 1930. "The earthquake at Cumaná, Venezuela," *Bulletin of the Seismological Society of America*, **20**: 1–10.

Page, W. D. 1986. *Seismic Geology and Seismicity of Northwestern Colombia.* San Francisco: Woodward-Clyde Consultants.

Papazachos, B. C., C. A. Papaioannou, C. B. Papazachos, and A. S. Savvaidis. 1997. *Atlas of Isoseismal Maps for Strong Shallow Earthquakes in Greece and Surrounding Area (426 BC–1995).* Thessaloniki: University of Thessaloniki.

Pasztory, E. 1997. *Teotihuacán.* Norman: University of Oklahoma Press.

Pearthree, P. A. and S. S. Calvo. 1987. "The Santa Rita fault zone: evidence for large magnitude earthquakes with very long recurrence intervals," *Bulletin of the Seismological Society of America*, **77**: 97–116.

Peet, S. D. 1903. "Ruined cities in Peru," *American Antiquarian and Oriental Journal*, **25**: 151–174.

Pelton, J. R., C. W. Meissner, and K. D. Smith. 1984. "Eyewitness account of normal surface faulting," *Bulletin of the Seismological Society of America*, **74**: 1083–1089.

Peñafiel, A. 1885. *Catálogo alfabético de los nombres de lugar pertenecientes al idioma "Nahuatl," estudio jeroglífico.* México: Oficina Tipografia de la Secretaría de Fomento.

Penick, J. 1976. *The New Madrid Earthquake of 1811–1812.* Columbia: University of Missouri Press.

Pennington, W. D. 1981. "Subduction of the Eastern Panama Basin and the seismotectonics of northwestern South America," *Journal of Geophysical Research*, **86**: 10 753–10 770.

Peraldo, H. G. and P. W. Montero. 1994. *Los temblores del periodo colonial de Costa Rica*. Cartago: Editorial Tecnologica de Costa Rica.

Perrey, A. 1854. "Documents relatifs aux tremblements de terre au Chili," *Annales de la Société Imperiale d'Agriculture, d'Histoire Naturelle et des Arts Utiles de Lyon*, 1–206.

　1857. "Documents sur les tremblements de terre au Pérou, dans la Colombie et dans le bassin de l'Amazone," *Memoires Academie Royale des Sciences, des Lettres, et des Beaux-Arts de Belgique*, **7**: 1–135.

Peterson, M. D. and S. G. Wesnousky. 1994. "Fault slip rates and earthquake histories for active faults in southern California," *Bulletin of the Seismological Society of America*, **84**: 1608–1649.

Plafker, G. 1976. "Tectonic aspects of the Guatemala earthquake," *Science*, **193**: 1201–1208.

Plafker, G. and J. C. Savage. 1970. "Mechanism of the Chilean earthquakes of May 21 and 22, 1960," *Geological Society of America Bulletin*, **81**: 1001–1030.

Plafker, G. and S. Ward. 1992. Backarc thrust faulting and tectonic uplift along the Caribbean sea coast during the April 22, 1991, Costa Rica earthquakes," *Tectonics*, **11**: 709–718.

Plafker, G., G. E. Erickson, and J. F. Concha. 1971. "Geological aspects of the May 31, 1970 Peru earthquake," *Bulletin of the Seismological Society of America*, **61**: 543–578.

Poey y Aquirre, A. 1855. *Tableu chronologique des tremblements de terre resentis à l'Ile de Cuba de 1551 à 1855*. Paris: A. Bertrand.

　1858. "Catalogue chronologique des tremblements de terre ressentis dans les Indes-Occidentales de 1530 à 1857," *Annuaire Société Météorologique de France*, **5**: 75–127, 227–252.

Pollock, H. E. D. 1937. *The Casa Redonda at Chichen Itza, Yucatán*, Carnegie Institution of Washington Publication 456. Washington, DC.

Polo, J. T. 1898. "Sinópsis de temblores del Perú," *Boletin de la Sociedad Geográfica de Lima*, **8**: 323–416.

　1899. "Sinópsis de temblores del Perú," *Boletin de la Sociedad Geográfica de Lima*, **9**: 15–95.

　1904. "Sinópsis de temblores del Perú," *Boletin de la Sociedad Geográfica de Lima*, **16**: 91–118.

Poole, L. and G. Poole. 1962. *Volcanoes in Action: Science and Legend*. New York: McGraw-Hill.

Potter, D. F. 1977. "Maya architecture of the central Yucatán peninsula, Mexico," *Middle American Research Institute, Tulane University*, Publication 44. New Orleans.

Prentice, C. S. and D. P. Schwartz. 1991. "Re-evaluation of 1906 surface faulting, geomorphic expression, and seismic hazard in the southern Santa Cruz mountains," *Bulletin of the Seismological Society of America*, **81**: 1424–1479.

Proskouriakoff, T. 1946. *An album of Maya architecture*, Carnegie Institution of Washington Publication 558. Washington, DC.

Rabiela, T. R., Zevallos, J. M. P., and V. G. Acosta (eds.). 1987. *Y volvio a temblar cronologia de los sismos en Mexico (de 1 pedernal a 1821)*. México, DF: Cuadernos de la Casa Chata 135.

Radin, P. 1920. "The sources and authenticity of the history of the ancient Mexicans," *University of California Publications in American Archaeology and Ethnology*, **17**: 1–50.

Ramirez, J. E. 1933. "Earthquake history of Colombia," *Bulletin of the Seismological Society of America*, **23**: 13–22.

    1948. "The Pasto, Colombia earthquake of July 14, 1947," *Bulletin of the Seismological Society of America*, **38**: 247–256.

    1951. "El gran terremoto ecuatoriano de Pelileo," *Revista de la Academia Colombiana de Ciencias Exactas Fisicas y Naturales, Bogata*, **8**: 126–136.

    1971. "The destruction of Bahia Solano, Colombia, on September 26, 1970 and the rejuvenation of a fault," *Bulletin of the Seismological Society of America*, **61**: 1041–1049.

    1974. *Historia de los terremotos en Colombia*. Bogota: Instituto Geográfico "Agustin Codazzi."

Ramos, A. E. 1946. "Algunes ruinas prehispánicas en Quintana Roo," *Boletin de la Sociedad Mexicana de Geográfia y Estadistica*, **61**: 513–628.

Rapp, G. 1986. "Assessing archaeological evidence for seismic catastrophies," *Geoarchaeology*, **1**: 365–379.

Rasmussen, N. 1967. "Washington State earthquakes 1840 through 1965," *Bulletin of the Seismological Society of America*, **57**: 463–476.

Recinos, A. 1950. *Popol Vuh: The Sacred Book of the Ancient Quiché Maya*. Norman: University of Oklahoma Press.

Recinos, A., D. Goetz, and D. Chonay (trans.). 1953. *The Annals of the Cakchiquels*. Norman: University of Oklahoma Press.

Reid, H. F. 1911. "Remarkable earthquakes in central New Mexico in 1906 and 1907," *Bulletin of the Seismological Society of America*, **1**: 10–16.

Reid, H. F. and S. Taber. 1919. "The Porto Rico earthquakes of October–November, 1918," *Bulletin of the Seismological Society of America*, **9**: 95–127.

    1920. "The Virgin Islands earthquakes of 1867–1868," *Bulletin of the Seismological Society of America*, **10**: 9–30.

Reiter, L. 1990. *Earthquake Hazard Analysis: Issues and Insights*. New York: Columbia University Press.

Restall, M. 1998. *Maya Conquistador*. Boston: Beacon Press.

Richards, H. G. and W. Broeker. 1963. "Emerged Holocene South American shorelines," *Science*, **141**: 1044–1045.

Richins, W. D., J. C. Pechmann, R. B. Smith, *et al.*, 1987. "The 1983 Borah Peak, Idaho earthquake and its aftershocks," *Bulletin of the Seismological Society of America*, **77**: 694–723.

Richter, C. F. 1935. "An instrumental earthquake magnitude scale," *Bulletin of the Seismological Society of America*, **25**: 1–32.

1958a. *Elementary Seismology*. San Francisco: W. H. Freeman and Co.

1958b. *The Magnitude Scale: its use and misuse. Mimeographed Notes*. California: Seismological Laboratory, California Institute of Technology.

1959. "Seismic regionalization," *Bulletin of the Seismological Society of America*, **49**: 123–162.

1970. "Magnitude of the Inglewood, California earthquake of June 21, 1920," *Bulletin of the Seismological Society of America*, **60**: 647–649.

Richter, C. F., C. R. Allen, and J. M. Nordquist. 1958. "The Desert Hot Springs earthquakes and their tectonic environment," *Bulletin of the Seismological Society of America*, **48**: 315–337.

Ricketson, O. G., Jr. and E. B. Ricketson. 1937. *Uaxactun, Guatemala, Group E, 1926–31*. Carnegie Institution of Washington Publication 477. Washington, DC.

Rivera, J. 1983. *Sismos en Ayacucho*. Peru: Universidad Nacional de San Cristobal de Huamanga.

Robertson, T. A. (trans.). 1968. *My Life Among the Savage Nations of New Spain*. Los Angeles: The Ward Ritchie Press.

Robinson, A. and P. Talwani. 1983. "Building damage at Charleston, South Carolina, associated with the 1886 earthquake," *Bulletin of the Seismological Society of America*, **73**: 633–652.

Robson, G. R. 1964. "An earthquake catalog for the eastern Caribbean," *Bulletin of the Seismological Society of America*, **54**: 785–832.

Rockstroh, E. 1883. "Temblores y erupciones en Centro-America," *Revista del Observatorio Central de Guatemala*, **1**: 21–42.

Rod, E. 1956. "Strike-slip faults of northern Venezuela," *American Association of Petroleum Geologists Bulletin*, **40**: 457–476.

Rodriguez, R. G. 1990. *Chan Chan*. Lima, Peru: Consejo Nacional de Ciencia y Techología, Instituto Indigenista Peruano.

Rogers, A. M., S. C. Harmsen, W. J. Carr, and W. Spence. 1983. "Southern Great Basin seismological data report for 1981 and preliminary data analysis," *US Geological Survey Open File Report* 83-669: 1–240.

Rogers, A. M., T. J. Walsh, W. J. Kickelman, and G. R. Priest. 1996. *Assessing earthquake hazards and reducing risk in the Pacific Northwest*, US Geological Survey Professional Paper 1560.

Rojas, W., H. Bungum, and C. Lindholm. 1993. "Historical and recent earthquakes in Central America," *Revista geológica de América Central*, **16**: 5–21.

Roosevelt, C. V. S. 1935. "Ancient civilizations of the Santa Valley and Chavín," *Geographical Review*, **25**: 21–42.

Rostworowski de Diez Canseco, M. 1978–80. "Guarco y Lunaguna. Dos señoríos prehispánicos de la costa sur central del Perú," *Revista del Museo Nacional*, **44**: 153–214.

Roys, R. L. 1934. *The Engineering Knowledge of the Maya*. Carnegie Institution of Washington Publication 436, Contribution 6. Washington, DC.

1967. *The Book of Chilam Balam of Chumayel*. Norman: University of Oklahoma Press.

Rudolph, E. and S. Szirtes. 1911–12. *Das Kolumbianische Erdbeben am 31 Januar 1906*, 3 vols. Leipzig: Verlag von Wilhelm Engelmann.

Rudolph, E. and E. Tams. 1907. *Seismogramme des Nordpazifischen und Südamerikanischen Erdbebens am 16 August 1906: Begleitworte und Erläuterungen.* Strasburg.

Ruppert, K. 1925. "Report on secondary constructions in the Court of Columns," *Carnegie Institution of Washington Year Book* 24: 269–270.

1931. *Temple of the Wall Panels, Chichen Itza.* Carnegie Institution of Washington Publication 403. Washington, DC.

1943. *The Mercado, Chichen Itza, Yucatán, Mexico.* Carnegie Institution of Washington Publication 546, Contribution 43. Washington, DC.

1952. *Chichen Itza: architectural notes and plans.* Carnegie Institution of Washington Publication 595. Washington, DC.

Ruppert, K., E. M. Shook, A. L. Smith, and R. E. Smith. 1954. "Chichen Itza, Dzibiac, and Balam Canche, Yucatán," *Carnegie Institution of Washington Year Book* 53: 240–247.

Russell, F. 1908. *The Pima Indians.* (Extract from the 26th Annual Report of the Bureau of American Ethnology). Washington: Government Printing Office, 3–389.

Rutten, L. and B. van Raadshooven. 1940. "On earthquake epicenters and earthquake shocks between 1913 and 1938 in the region between 0° and 30° N and 56° and 120° W," *Verhandelingen der Nederlandsche Akademie van Wetenschappen, Afdeeling Natuurkunde (Tweede Sectie)*, **39**: 1–44.

Ryall, A. 1962. "The Hebgen Lake, Montana, earthquake of August 18, 1959: P waves," *Bulletin of the Seismological Society of America*, **52**: 235–271.

Sabloff, J. E. 1990. *The New Archaeology and the Ancient Maya.* New York: W. H. Freeman.

St. Amand, P. 1961. *Los Terremotos de Mayo-Chile 1960.* Technical Article 14. China Lake, California: US Naval Ordnance Test Station.

Sampson, F. A. 1913. "The New Madrid and other earthquakes of Missouri," *Bulletin of the Seismological Society of America*, **3**: 57–71.

Sangawa, A. 1992. *Earthquake Archeology* (in Japanese). Tokyo: Chuko Shinsho.

Sansores, M. C. 1956. *Chi Cheen Itsa Archaeological Paradise of America.* Merida: Privately Printed.

Sapper, C. 1925. *Los volcanes de la América Central.* Halle (Saale): Verlag von Max Niemeyer.

Sapper, K. 1902. "Das Erdbeben in Guatemala vom 18 April 1902," *Petermanns geographische Mittheilungen*, **48**: 193–195.

Satake, K., K. Shimazaki, Y. Tsuji, and K. Ueda. 1996. "Time and size of a giant earthquake in Cascadia inferred from Japanese tsunami records of January 1700," *Nature*, **379**: 246–249.

Satterthwaite, L. 1958. "The problem of abnormal stela placements at Tikal and elsewhere," *Tikal Reports, Museum Monographs*, **3**. Philadelphia: The University Museum.

1961. "The mounds and monuments at Xutilha, Petén, Guatemala," *Tikal Reports, Museum Monographs*, 9. Philadelphia: The University Museum.

Saucier, R. T. 1991. "Geoarchaeological evidence of strong prehistoric earthquakes in the New Madrid (Missouri) seismic zone, *Geology*, **19**: 296–298.

Saville, M. H. 1892. "Explorations on the Main Structure of Copán, Honduras," *Proceedings of the American Association for the Advancement of Science*, **41**: 271–275.

1918a. "The discovery of Yucatán in 1517 by Francisco Hernandez de Córdoba," *Geographical Review*, **6**: 436–448.

1918b. "The Guatemala earthquake of December 1917 and January 1918," *Geographical Review*, **5**: 459–469.

Schele, L. 1982. *Maya Glyphs: The Verbs.* Austin: University of Texas Press.

Schellhas, P. 1904. "Representation of deities of the Maya manuscripts," *Papers of the Peabody Museum of American Archaeology and Ethnology, Harvard Museum*, **4**: 1–46.

Scherer, J. 1912a. "Great earthquakes in the island of Haiti," *Bulletin of the Seismological Society of America*, **2**: 161–180.

1912b. "Notes on remarkable earthquake sounds in Haiti," *Bulletin of the Seismological Society of America*, **2**: 230–232.

Scherzer, K. 1855. "Ein Besuch bei der Ruinen von Quiriguá," *Sitzungsberichte der Klasse Akademie der Wissenschaften*, **16**, Vienna.

Schneider, J. F., W. D. Pennington, and R. P. Meyer. 1987. "Microseismicity and focal mechanisms of the intermediate-depth Bucaramanga nest, Colombia," *Journal of Geophysical Research*, **92**: 13 913–13 926.

Scholz, C. H. 1982. "Scaling laws for large earthquakes: consequences for physical models," *Bulletin of the Seismological Society of America*, **72**: 1–14.

1990. *The Mechanics of Earthquakes and Faulting.* Cambridge: Cambridge University Press.

1997. "Size distributions for large and small earthquakes," *Bulletin of the Seismological Society of America*, **87**: 1074–1077.

Schuster, R. L. and W. Murphy. 1996. "Structural damage, ground failure, and hydrologic effects of the magnitude ($M_w$) 5.9 Draney Peak, Idaho earthquake of February 3, 1994," *Seismological Research Letters*, **67**: 20–29.

Schwartz, D. P. 1988. "Paleoseismicity and neotectonics of the Cordillera Blanca fault zone, northern Peruvian Andes," *Journal of Geophysical Research*, **89**: 5681–5698.

Schwartz, D. P., L. S. Cluff, and T. W. Donnelly. 1979. "Quaternary faulting along the Caribbean–North American plate boundary in Central America," *Tectonophysics*, **52**: 431–445.

Schweig, E. S., M. Tuttle, Y. Li *et al.*, 1993. "Evidence for recurrent strong earthquake shaking in the past 5000 years, New Madrid region, Central US," *EOS, Transactions of the American Geophysical Union*, **74**: 438.

Seismological notes. *Bulletin of the Seismological Society of America*, **1–84**.

Seler, E. 1902–23. *Gesammelte Abhandlungen zur Amerikanischen Sprach und Alterthumskunde*, 5 vols. Berlin: Ascher, Behrend.

1915. *Beobachtungen und Studien in den Ruinen von Palenque.* Berlin: Verlag der Königlichen Akademie der Wissenschaften.

1990. *Collected Works in Mesoamerican Linguistics and Archaeology*, 5 vols. (translation). Culver City, California: Labyrinthos.

Shapiro, N. M., S. K. Singh, and J. Pacheco. 1998. "A fast and simple diagnostic method for identifying tsunamigenic earthquakes," *Geophysical Research Letters*, **25**: 3911–3914.

Sharer, R. J. 1969. "Chalchuapa: investigations at a highland Maya ceremonial center," *Expedition*, **11**(2): 36–38.

1974. "The prehistory of the southeastern Maya periphery," *Current Anthropology*, **15**(2): 165–187.

1978. "Archaeology and history at Quiriguá, Guatemala," *Journal of Field Archaeology*, **5**: 51–70.

1990. *Quiriguá: A Classic Maya Center and Its Sculptures*. Durham: Carolina Academic Press.

1994. *The Ancient Maya*. Stanford: Stanford University Press.

1996. *Daily Life in Maya Civilization*. Westport: Greenwood Press.

Sheets, P. D. 1971a. "An ancient natural disaster," *Expedition*, **14**(1): 24–31.

1971b. "Maya recovery from volcanic disasters," *Archaeology*, **32**(3): 32–42.

1979. *Volcanic Activity and Human Ecology*. New York: Academic Press.

(ed.). 1983. *Archaeology and Volcanism in Central America: The Zapotitlan Valley of El Salvador*. Austin: University of Texas Press.

1992. *The Ceren Site: A Prehistoric Village Buried by Volcanic Ash in Central America*. Fort Worth: Harcourt Brace Jovanovich College Publishers.

Shepard, E. M. 1905. "The New Madrid earthquake," *Journal of Geology*, **23**: 45–62.

Shi, B., A. Anooshehpoor, Y. Zeng, and J. N. Brune. 1996. "Rocking and overturning of precariously balanced rocks by earthquakes," *Bulletin of the Seismological Society of America*, **86**: 1364–1371.

Shimada, I., C. B. Schaaf, L. G. Thompson, and E. Mosley-Thompson. 1991. "Cultural impacts of severe droughts in the prehistoric Andes: application of a 1500-year ice core precipitation record," *World Archaeology*, **22**(3): 247–270.

Shippee, R. 1932. "The 'Great Wall of Peru' and other aerial photographic studies by the Shippee–Johnson Peruvian expedition," *Geographical Review*, **22**: 1–29.

1933. "Air adventures in Peru," *National Geographic*, **63**: 80–120.

Shook, E. M. 1952. "Lugares arqueologicos del altiplano central de Guatemala," *Antropologia e Historia de Guatemala*, **4**(2): 3–40.

Sibol, M. S., G. A. Bollinger, and J. B. Birch. 1987. "Estimation of magnitudes in central and eastern North America using intensity and felt area," *Bulletin of the Seismological Society of America*, **77**: 1635–1654.

Sidrys, R. V., C. M. Krowne, and H. B. Nicholson. 1975. "A lowland Maya Long Count/Gregorian conversion computer program," *American Antiquity*, **40**: 337–344.

Sieberg, A. 1930. "Los terremotos en el Perú," in *Geologia del Perú*, ed. G. Steinman. Heidelberg: Karl Winter, 406–421.

1932. "Erdbebengeographie," in *Handbuch der Geophysik*, Vol. 4. Berlin: Gebrüder Borntraeger, 687–1005.

Sieh, K. E. 1977. "A study of late Holocene displacement history along the south–central reach of the San Andreas fault," PhD Dissertation, Stanford University.

  1978a. "Prehistoric large earthquakes produced by slip on the San Andreas fault at Pallett Creek, California," *Journal of Geophysical Research*, **83**: 3907–3939.

  1978b. "Slip along the San Andreas fault associated with the great 1857 earthquake," *Bulletin of the Seismological Society of America*, **68**: 1421–1428.

  1984. "Lateral offsets and revised dates of large prehistoric earthquakes at Pallett Creek, southern California," *Journal of Geophysical Research*, **89**: 7641–7670.

  1996. "The repitition of large-earthquake ruptures," *Proceedings of the National Academy of Sciences of the USA*, **93**: 3764–3771.

Sieh, K. E. and R. H. Jahns. 1984. "Holocene activity of the San Andreas fault at Wallace Creek, California," *Geological Society of America Bulletin*, **95**: 883–896.

Sieh, K., M. Stuiver, and D. Brillinger. 1989. "A more precise chronology of earthquakes produced by the San Andreas fault in southern California," *Journal of Geophysical Research*, **94**: 603–623.

Sievers, H. A., G. Villegas, and G. Barros. 1963. "The seismic sea wave of 22 May 1960 along the Chilean coast," *Bulletin of the Seismological Society of America*, **53**: 1125–1190.

Silgado, E. F. 1951. "The Ancash, Peru earthquake of November 10, 1946," *Bulletin of the Seismological Society of America*, **41**: 83–100.

  1968. "Historia de los sismos más notables ocurridos en el Perú (1515–1960)," *Boletin Bibliográfico de Geografia Oceanografia Americanas*, **4**: 191–241.

  1973. "Historia de los sismos más notables ocurridos en el Perú (1555–1970)," *Geofisica Panamerica*, **2**(1): 179–243.

  1992. *Investigacion de sismicidad historica en la America de sur en los siglos XVI, XVII, XVIII, XVIX*. Lima: Consejo Nacional de Ciencia y Tecnología.

Silverman, H. 1993. *Cahuachi in the Ancient Nasca World*. Iowa City: University of Iowa Press.

Simkin, T., L. Siebert, L. McCleeland, *et al.*, 1981. *Volcanoes of the World*. Stroudsburg: Hutchinson Ross Publishing Company.

Singh, S. K. and J. Havskov. 1980. "On moment-magnitude scale," *Bulletin of the Seismological Society of America*, **70**: 379–383.

Singh, S. K. and J. Pacheco. 1994. "Magnitude of Mexican earthquakes," *Geofisica International*, **33**: 189–198.

Singh, S. K., M. Reichle, and J. Havskov. 1980. "Magnitude and epicenter estimations of Mexican earthquakes from isoseismic maps," *Geofísica Internacional (México)*, **19**: 269–284.

Singh, S. K., L. Astiz, and J. Haskov. 1981. "Seismic gaps and recurrence periods of large earthquakes along the Mexican subduction zone: a reexamination," *Bulletin of the Seismological Society of America*, **71**: 827–843.

Singh, S. K., M. Rodriguez, and L. Esteva. 1983. "Statistics of small earthquakes and frequency of occurrence of large earthquakes along the Mexican subduction zone," *Bulletin of the Seismological Society of America*, **73**: 1779–1796.

Singh, S. K., M. Rodriguez, and J. M. Espindola. 1984. "A catalog of shallow earthquakes of Mexico from 1900–1981," *Bulletin of the Seismological Society of America*, **74**: 267–274.

Singh, S. K., L. Ponce, and S. P. Nishenko. 1985. "The great Jalisco, Mexico earthquakes of 1932: subduction of the Rivera plate," *Bulletin of the Seismological Society of America*, **75**: 1301–1313.

Singh, S. K., L. Astiz, and J. Havskov. 1987. "Seismic gaps and recurrence periods of large earthquakes along the Mexican subduction zone: a reexamination," *Bulletin of the Seismological Society of America*, **71**: 827–843.

Singh, S. K., M. Ordaz, J. G. Anderson, *et al.*, 1989. "Analysis of near-source strong-motion recordings along the Mexican subduction zone," *Bulletin of the Seismological Society of America*, **79**: 1697–1717.

Singh, S. K., M. Ordaz, J. F. Pacheco, *et al.*, 1999. "A preliminary report on the Tehuacán Mexico earthquake of June 15, 1999 ($M_w = 7.0$)." *Seismological Research Letters*, **70**: 489–504.

Singh, S. K., M. Ordaz, and L. E. Pérez-Rocha. 1996. "The great Mexican earthquake of 19 June 1858: expected ground motions and damage in Mexico City from a similar future event," *Bulletin of the Seismological Society of America*, **86**: 1655–1666.

Singh, S. K., M. Ordaz, L. Alcántara, *et al.*, 2000. "The Oaxaca earthquake of 30 September 1999 ($M_w = 7.5$): a normal-faulting event in the subducted Cocos plate," *Seismological Research Letters*, **71**: 67–78.

Slemmons, D. B., A. E. Jones, and J. I. Gimlett. 1965. "Catalog of Nevada earthquakes, 1852–1960," *Bulletin of the Seismological Society of America*, **55**: 519–566.

Smith, A. L. 1929. "Reports on excavation of the Stelae at Uaxactun and on the map of the environs of Uaxactun," *Carnegie Institution of Washington Year Book* 28: 323–329.

1940. "The Corbelled Arch in the New World," in *The Maya and their Neighbors,* ed. C. L. Hay *et al.* New York: Appleton-Century Co., 202–221.

1950. *Uaxactun, Guatemala: Expeditions of 1931–1937.* Carnegie Institution of Washington Publication 588. Washington, DC.

1972. "Excavations at Altar de Sacrificios," *Papers of the Peabody Museum of Archaeology and Ethnology, Harvard University,* **62**(2): 1–282.

1982. "Excavations at Seibal, major architecture and caches," *Memoirs of the Peabody Museum of Archaeology and Ethnology, Harvard University,* **15**(1): 1–263.

Smith, D. E. and F. J. Teggart (eds.). 1910. "Diary of Gaspar de Portola during the California expedition of 1769–1770," *Publications of the Academy of Pacific Coast History*, **1**: 33–59.

Smith, R. B. and M. L. Sbar. 1974. "Contemporary tectonics and seismicity of the western United States with emphasis on the intermountain seismic belt," *Geological Society of America Bulletin*, **85**: 1205–1218.

Smith, R. S. U. 1979. "Holocene offset and seismicity along the Panamint Valley fault zone, Western Basin-and-Range Province, California," *Tectonophysics*, **52**: 411–415.

Smith, W. E. T. 1962. "Earthquakes of eastern Canada and adjacent areas 1534–1927," *Publications of the Dominion Observatory, Ottawa*, **26**: 271–301.

 1966. "Earthquakes of eastern Canada and adjacent areas, 1928–1959," *Publications of the Dominion Observatory, Ottawa*, **32**: 87–121.

Snow, D. R. 1994. *The Iroquois*. Oxford: Blackwell.

Solis, E. C., R. M. S. Roman, M. Reyes, and F. C. Ruiz. 1987. *Vocabulario Nautl-Español de la Huasteca*. Mexico: Gobierno del Estado de Hidalgo.

Sornette, D. and L. Knopoff. 1997. "The paradox of the expected time until the next earthquake," *Bulletin of the Seismological Society of America*, **87**: 789–798.

Spence, L. 1923. *The Gods of Mexico*. London: T. Fisher Unwin Ltd.

Spier, L. 1933. *Yuman Tribes of the Gila River*. Chicago: University of Chicago Press.

Spinden, H. J. 1913. "A study of Maya art," *Memoirs of the Peabody Museum of Archaeology and Ethnology, Harvard University*, **6**: 1–337.

Spofford, C. M. 1911. "Earthquake effects on structures at Cartago, Costa Rica," *Association of Engineering Societies*, **46**: 63–80.

Staunton, W. F. 1918. "Effects of an earthquake in a mine at Tombstone," *Bulletin of the Seismological Society of America*, **8**: 25–27.

Stein, R. S. and S. E. Barrientos. 1985. "The 1983 Borah Peak, Idaho earthquake: geodetic evidence for deep rupture on a planar fault," *US Geological Survey Open File Report* 85-290: 459–484.

Steinbrugge, K. V. and W. K. Cloud. 1962. "Epicentral intensities and damage in the Hebgen Lake, Montana earthquake of August 17, 1959," *Bulletin of the Seismological Society of America*, **52**: 181–234.

Steinbrugge, K. V., W. K. Cloud, and N. H. Scott. 1970. *The Santa Rosa, California earthquakes of October 1, 1969*. US Department of Commerce.

Stephens, J. L. 1841. *Incidents of Travel in Central America, Chiapas and Yucatán*, 2 vols. London: John Murray.

 J. L. 1843. *Incidents of Travel in Yucatán*, 2 vols. London: John Murray.

Stevens, A. E. 1980. "Reexamination of some larger La Malbaie, Quebec earthquakes (1924–1978)," *Bulletin of the Seismological Society of America*, **70**: 529–557.

 1995. "Eastern North American earthquakes prior to 1660," *Bulletin of the Seismological Society of America*, **85**: 1398–1415.

Stewart, I. S. and V. A. Buck. 2001. "Earthquake archaeology: a logical approach?," *EOS, Transactions of the American Geophysical Union*, **82**(47): 31.

Stiros, S. C. 1996. "Identification of earthquakes from archaeological data: methodology, criteria and limitations," in *Archaeoseismology*, eds. S. Stiros and R. E. Jones, Fitch Laboratory Occasional Paper 7. Oxford: British School at Athens, 129–152.

Stone, D. 1941. "Archaeology of the north coast of Honduras," *Memoirs of the Peabody Museum of Archaeology and Ethnology, Harvard University*, **9**(1), 1–103.

Stover, C. W. 1985. "The Borah Peak, Idaho earthquake of October 28, 1983: isoseismal map and intensity distribution," *Earthquake Spectra*, **2**: 11–16.

Stover, C. W. and J. L. Coffman. 1993. *Seismicity of the United States, 1568–1989 (Revised)*. US Geological Survey Professional Paper 1527.

Street, R. L. and A. Lacroix. 1979. "An empirical study of New England seismicity: 1727–1977," *Bulletin of the Seismological Society of America*, **69**: 159–176.

Stromsvík, G. 1935. "Copán," *Carnegie Institution of Washington Year Book* 34: 118–119.

　　1941. *Substela caches and stela foundations at Copán and Quiriguá.* Carnegie Institution of Washington Publication 528, Contributions to American Anthropology and History 37. Washington, DC.

　　1947. *Guide book to the ruins of Copán.* Carnegie Institution of Washington Publication 577. Washington, DC.

Suárez, G., P. Molnar, and B. C. Burchfiel. 1983. "Seismicity, fault plane solutions, depth of faulting, and active tectonics of the Andes of Peru, Ecuador, and southern Colombia," *Journal of Geophysical Research*, **88**: 10 403–10 428.

Sugawara, M. 1987. "Notas sobre los sismos Mexicanos en el siglo XVI," *Históricas*, **22**: 3–17.

Sullivan, W. 1996. *The Secret of the Incas.* New York: Crown Publishers, Inc.

Sultan, D. I. 1931. "The Managua earthquake of 1931," *Military Engineer*, **23**: 354–361.

Sumner, J. R. 1977. "The Sonora earthquake of 1887," *Bulletin of the Seismological Society of America*, **67**: 1219–1223.

Sutch, P. L. 1979. "Historic Seismicity of Honduras, 1539–1978," PhD thesis, Stanford University.

　　1981. "Estimated intensities and probable tectonic sources of historic (pre-1898), Honduras earthquakes," *Bulletin of the Seismological Society of America*, **71**: 865–881.

Suter, M. 2001. "The historical seismicity of northeastern Sonora and northwestern Chichuahua (28–32° N, 106–111° W)," *Journal of South American Earth Sciences*, **14**: 521–532.

Suter, M. and J. Contreras. 2002. "Active tectonics of northeastern Sonora, Mexico (southern Basin and Range Province) and the 3 May 1887 $M_w$ 7.4 earthquake," *Bulletin of the Seismological Society of America*, **92**: 581–589.

Suter, M., M. Carillo-Martinez, and O. Quintero-Legoretta. 1996. "Macroseismic study of shallow earthquakes in the central and eastern parts of the Trans-Mexican Volcanic Belt, Mexico," *Bulletin of the Seismological Society of America*, **86**: 1952–1963.

Swan, J. G. 1857. *The Northwest Coast.* New York: Harper and Brothers.

Sykes, L. R. and M. Ewing. 1965. "The seismicity of the Caribbean Region," *Journal of Geophysical Research*, **70**: 5065–5074.

Symposium. 1925. "The Santa Barbara earthquake," *Bulletin of the Seismological Society of America*, **15**: 251–333.

Taber, S. 1914. "Seismic activity in the Atlantic coastal plain near Charleston, South Carolina," *Bulletin of the Seismological Society of America*, **4**: 108–160.

　　1920a. "Jamaica earthquakes and the Bartlett Trough," *Bulletin of the Seismological Society of America*, **10**: 55–89.

　　1920b. "The Inglewood Earthquake in southern California, June 21, 1920," *Bulletin of the Seismological Society of America*, **10**: 129–145.

1922. "The seismic belt in the Greater Antilles," *Bulletin of the Seismological Society of America*, **12**: 198–219.

Takeo, M. 1998. "Ground rotational motions recorded in near-source region," *Geophysical Research Letters*, **25**: 789–792.

Takeo, M. and H. M. Ito. 1997. "What can be learned from rotational motions excited by earthquakes?" *Geophysical Journal International*, **129**: 319–329.

Talley, H. C., Jr. and W. K. Cloud. 1962. *United States Earthquakes 1960*. US Department of Commerce Coast and Geodetic Survey.

Talwani, P. 1977. "An intensity survey of the April 28, 1975 Summerville, South Carolina earthquake," *Bulletin of the Seismological Society of America*, **67**: 547–549.

1999. "Fault geometry and earthquakes in continental interiors," *Tectonophysics*, **305**: 371–379.

Talwani, P. and J. Cox. 1985. "Paleoseismic evidence for recurrence of earthquakes near Charleston, South Carolina," *Science*, **229**: 379–381.

Talwani, P. and W. T. Schaeffer. 2001. "Recurrence rates for large earthquakes in the South Carolina coastal plain based on liquefaction data," *Journal of Geophysical Research*, **106**: 6621–6642.

Talwani, P. and N. Sharma. 1999. "Reevaluation of the magnitudes of three destructive aftershocks of the 1886 Charleston earthquake," *Seismological Research Letters*, **70**: 360–367.

Taracena Flores, A. 1970. *Los terremotos de Guatemala*. Guatemala City: Tipografía Nacional.

Tarr, A. C., P. Talwani, S. Rhea, D. Carver, and D. Amick. 1981. "Results of recent South Carolina seismological studies," *Bulletin of the Seismological Society of America*, **71**: 1883–1902.

Tasdemiroglu, M. 1971. "The 1970 Gediz earthquakes in western Anatolia, Turkey," *Bulletin of the Seismological Society of America*, **61**: 1507–1527.

Tedlock, D. 1985. *Popol Vuh: The Mayan Book of the Dawn of Life*. New York: Simon and Schuster.

Thatcher, W., G. R. Foulger, B. R. Julian, *et al.*, 1999. "Present-day deformation across the Basin and Range Province, western United States," *Science*, **283**: 1714–1718.

Thompson, E. H. 1904. "Archaeological researches of Yucatán," *Memoirs of the Peabody Museum of Archaeology and Ethnology, Harvard University*, 3(1): 1–30.

Thompson, J. E. S. 1939. *Excavations at San José, British Honduras*, Carnegie Institution of Washington Publication 506. Washington, DC.

1954. *The Rise and Fall of the Maya Civilization*. Norman: University of Oklahoma Press.

1970. *Maya History and Religion*. Norman: University of Oklahoma Press.

Thompson, J. E. S., H. E. D. Pollock, and J. Charlot. 1932. *A Preliminary Study of the Ruins of Coba, Quintana Roo, Mexico*. Carnegie Institution of Washington Publication 424. Washington, DC.

Thompson, L. G., E. Moseley-Thompson, J. F. K. Bolzen, and B. R. Koci. 1985. "A 1500-year record of tropical precipitation from the Quelccaya ice cap, Peru," *Science*, **229**: 971–973.

Thwaites, R. G. (trans.). 1896–1901. *The Jesuit Relations and Allied Documents: Travels and Explorations of the Jesuit Missionaries in New France, 1610–1791,* 73 vols. Cleveland: Burrows Brothers Company.

Tocher, D. 1956. "Movement on the Rainbow Mountain Fault," *Bulletin of the Seismological Society of America,* **46**: 10–14.

1962. "The Hebgen Lake, Montana earthquake of August 17, 1959, MST," *Bulletin of the Seismological Society of America,* **52**: 153–162.

Tocher, D., C. Romney, C. A. Whitten, *et al.,* 1957. "The Dixie Valley–Fairview Peak, Nevada earthquakes of December 16, 1954," *Bulletin of the Seismological Society of America,* **47**: 299–396.

Toiran, B. M. 2002. "The new Cuban seismograph network," *Seismological Research Letters,* **73**: 504–517.

Tomblin, J. M. and G. R. Robson. 1977. *A Catalogue of Felt Earthquakes for Jamaica, with References to Other Islands in the Greater Antilles.* Kingston: Mines and Geology Division, Ministery of Mining and Natural Resources.

Toppozada, T. R. 1975. "Earthquake magnitude as a function of intensity data in California and western Nevada," *Bulletin of the Seismological Society of America,* **65**: 1223–1238.

Toppozada, T. R. and G. Borchardt. 1998. "Re-evaluation of the 1836 'Hayward fault' and the 1838 San Andreas fault earthquakes," *Bulletin of the Seismological Society of America,* **88**: 140–159.

Toppozada, T. R., D. L. Parke, and C. T. Higgins. 1978. "Seismicity of California 1900–1931," *Special Report* 135, *California Division of Mines and Geology.*

Torquemada, F. J. de. 1969. *Monarquía Indiana,* 3 vols. México: Biblioteca Porrúa 41. Edicione Porrúa.

Totten, G. O. 1926. *Maya Architecture.* Washington, DC: The Maya Press.

Tournon, J. and G. Alvarado. 1997. *Carte géologique du Costa Rica.* Cartago: Editorial Tecnologica de Costa Rica.

Townley, S. D. and M. W. Allen. 1939. "Descriptive catalog of earthquakes of the Pacific coast of the United States 1769 to 1928," *Bulletin of the Seismological Society of America,* **29**: 1–297.

Tozzer, A. M. 1907. *A Comparative Study of the Mayas and the Lacandones.* New York: Archaeological Institute of America/MacMillan.

1941. "Landa's relación de las cosas de Yucatán," *Papers of the Peabody Museum of American Archaeology and Ethnology, Harvard University,* **18**: 1–394.

1957. "Chichen Itza and its Cenote of Sacrifice," *Memoirs of the Peabody Museum of Archaeology and Ethnology, Harvard University,* **11** and **12**.

Triep, E. and L. Sykes. 1997. "Frequency of occurrence of moderate to great earthquakes in intracontinental regions: implications for changes in stress, earthquake prediction, and hazards assessments," *Journal of Geophysical Research,* **102**: 9923–9948.

Trifunac, M. D. and A. G. Brady. 1975. "On the correlation of seismic intensity scales with the peaks of recorded ground motion," *Bulletin of the Seismological Society of America,* **65**: 139–162.

Trik, A. S. 1939. *Temple XXII at Copán*, Carnegie Institution of Washington Publication 509, Contributions to American Anthropology and History, 27. Washington, DC.

Tristan, J. F. 1916. "Costa Rica earthquake of February 27, 1916," *Bulletin of the Seismological Society of America*, **6**: 232–235.

Troll, C. (ed.). 1970. "Geo-ecology of the mountainous regions of the tropical Americas," *Proceedings of the UNESCO Mexico Symposium (1968)*, Vol. 9. Bonn: Ferdinand Dümmlers.

Tryggvason, E. and J. E. Lawson, Jr. 1970. "The intermediate earthquake source near Bucarmonga, Colombia," *Bulletin of the Seismological Society of America*, **60**: 269–276.

Tucker, G. 1956. *Tecumseh, Vision of Glory*. New York: Bobbs-Merrill Company, Inc.

Tuttle, M. P. and E. S. Schweig. 1995. "Archaeological and pedological evidence for large prehistoric earthquakes in the New Madrid seismic zone, central United States," *Geology*, **23**: 253–256.

  1996. "Recognizing the dating prehistoric liquefaction features: lessons learned in the New Madrid seismic zone, central United States," *Journal of Geophysical Research*, **103**: 6171–6178.

Tuttle, M. P. and L. R. Sykes. 1992. "Re-evaluation of several large historic earthquakes in the vicinity of Loma Prieta and peninsular segments of the San Andreas fault," *Bulletin of the Seismological Society of America*, **82**: 1802–1820.

Tuttle, M. P., R. H. Lafferty, M. J. Guccione, *et al.*, 1996. "Use of archaeology to date liquefaction features and seismic events in the New Madrid seismic zone, central United States," *Geoarchaeology*, **11**: 451–480.

Uhrhammer, R. 1996. "Seismic analysis of the Yosemite rock fall of July 10, 1996," *EOS, Transactions of the American Geophysical Union*, **77**: 508.

Umlauff, A. F. 1915. "La region sismica de Caraveli (Peru)," *Boletin de la Sociedad Geográfica de Lima*, **31**: 223–257.

Urbina, F. and H. Camacho. 1913. "La zona megaseismica Acambay–Tixmadeje, estado de Mexico," *Boletin del Instituto Geologico de Mexico*, **32**.

Vaillant, G. C. 1962. *Aztecs of Mexico*. Garden City, New York: Doubleday and Company, Inc.

Valencias, A. R. 1917. *Ultimos terremotos del Peru*. Lima: Imprenta de la Escuela de Ingenieros César Mesinas.

Van der Hilst, R. and P. Mann. 1994. "Tectonic implications of tomographic images of subducted lithosphere beneath northwestern South America," *Geology*, **22**: 451–454.

Vanek, J., A. Spicák, and V. Hanus. 2000. "Position of the disastrous 1999 Puebla earthquake in the seismotectonic pattern of Mexico," *Bulletin of the Seismological Society of America*, **90**: 786–789.

Velázquez, P. F. (trans.). 1975. *Anales de Cuauhtitlán*. Mexico City: UNAM, Universidad de Investigaciones Historicas.

Vered, M. and H. Striem. 1977. "A macroseismic study and the implications of structural damage of two recent major earthquakes in the Jordan Rift," *Bulletin of the Seismological Society of America*, **67**: 1607–1613.

Viitanen, W. 1973. "Folklore and fakelore of an earthquake," *Kentucky Folklore Record*, **19**: 99–111.

Villa Roiz, C. 1997. *Popocatépetl y la Mujer Prohibida.* Mexico: Plaza y Valdes.

Villacorta, C., J. A., and C. A. Villacorta. 1933. *Códices Mayas reproducidos y desarrollados.* Guatemala: Tipografía Nacional.

Vinson, G. L. 1962. "Upper cretaceous and tertiary stratigraphy of Guatemala," *American Association of Petroleum Geologists Bulletin*, **46**: 425–456.

Viquez, C. G. 1994. *Temblores, terremotos, inundaciones y erupciones volcanicas en Costa Rica 1608–1910.* Cartago: Editorial Tecnologica de Costa Rica.

Vitaliano, D. R. 1973. *Legends of the Earth: Their Geologic Origins.* Bloomington: Indiana University Press.

Vogt, E. Z. 1994. *Fieldwork among the Maya.* Albuquerque: University of New Mexico Press.

von Hake, C. A. and W. K. Cloud. 1966. *United States Earthquakes* 1964. US Department of Commerce.

Wadell, H. 1938. "Physical-geological features of Petén, Guatemala," in *The Inscriptions of Petén,* ed. S. G. Morley. Carnegie Institution of Washington Publication 437, Vol. 4: 336–348.

Waitz, P. and F. Urbina. 1919. "Los temblores de Guadalajara en 1912," *Instituto Geologico de Mexico Bulletin*, **19**: 1–81.

Wald, D. J., V. Quitoriano, T. H. Heaton, and H. Kanamori. 1999. "Relationships between peak ground acceleration, peak ground velocity and Modified Mercalli intensity in California." *Earthquake Spectra*, **15**: 557–564.

Wald, L. A., D. D. Given, J. Mori, L. M. Jones, and L. K. Hutton. 1990a. "The Southern California Network Bulletin January–December 1988," *US Geological Survey Open File Report* 90-499: 1–45.

Wald, L. A., D. D. Given, L. M. Jones, and L. K. Hutton. 1990b. "The Southern California Network Bulletin January–December 1989," *US Geological Survey Open File Report* 90-483: 1–22.

Wald, L. A., L. K. Hutton, J. Mori, D. D. Given, and L. M. Jones. 1991. "The Southern California Network Bulletin January–December 1990," *US Geological Survey Open File Report* 91-255: 1–45.

Wald, L. A., L. K. Hutton, L. M. Jones, *et al.*, 1992. "The Southern California Network Bulletin January–December 1991," *US Geological Survey Open File Report* 92-335: 1–49.

Wald, L. A., K. Watts, J. Mori, and K. Douglass. 1993. "The Southern California Network Bulletin January–December 1992," *US Geological Survey Open File Report* 93-227: 1–50.

Wald, L. A., S. Perry-Huston, and D. D. Givens. 1994. "The Southern California Network Bulletin January–December 1993," *US Geological Survey Open File Report* 94-199: 1–55.

Wald, L. A., K. Hafner, and S. Bryant. 1995. "The Southern California Network Bulletin January–December 1994," *US Geological Survey Open File Report* 95-204: 1–47.

Wald, L. A. and E. Hauksson. 1996. "The Southern California Network Bulletin January–December 1995," *US Geological Survey Open File Report* 96-29: 1–29.

Wald, L. A., J. Mori, and D. J. Wald. 1997. "The Southern California Network Bulletin January–December 1996," *US Geological Survey Open File Report* 97-133: 1–25.

Wallace, R. E. 1984. "Eyewitness account of surface faulting during the earthquake of 28 October 1983, Borah Peak, Idaho," *Bulletin of the Seismological Society of America*, **74**: 1091–1094.

Wallace, R. E., M. G. Bonilla, and H. A. Villalobos. 1984. *Faulting Related to the 1915 Earthquakes in Pleasant Valley, Nevada*. US Geological Survey Professional Paper 1274-A, B.

Wallace, W. J. and E. S. Taylor. 1955. "Early man in Death Valley," *Archaeology*, **8**(2): 88–92.

Ward, W. C., A. E. Weidie, and W. Beck. 1985. *Geology and Hydrogeology of the Yucatán and Quaternary Geology of Northeastern Yucatán Peninsula*. New Orleans: New Orleans Geological Society.

Wauchope, R. 1938. *Modern Maya Houses: A Study of their Archaeological Significance*. Carnegie Institution of Washington Publication 502. Washington, DC.

Weber, J., S. Stein, and J. Engeln. 1998. "Estimation of intraplate strain accumulation in the New Madrid seismic zone from repeat GPS surveys," *Tectonics*, **17**: 250–266.

Webster, D. 2002. *The Fall of the Ancient Maya*. London: Thames and Hudson.

Weeks, J. M. *Maya Civilization*. New York: Garland Publishing, Inc.

Weichert, D. 1994. "Omak Rock and the 1872 Pacific Northwest earthquake," *Bulletin of the Seismological Society of America*, **84**: 444–450.

Weidie, A. E. (comp.). 1967. *Yucatán Field Trip Guidebook, 1967, Annual Meeting of the Geological Society of America*. New Orleans: New Orleans Geological Society.

Weischet, W. 1963. "Further observation of geological and geomorphic changes resulting from the catastrophic earthquake of May 1960, in Chile," *Bulletin of the Seismological Society of America*, **53**: 1237–1257.

Weitzel, R. B. 1944. "Yucatecan chronological systems," *American Antiquity*, **13**: 53–58.

Wells, D. L. and K. J. Coppersmith. 1994. "New empirical relationships among magnitude, rupture length, rupture width, rupture area, and surface displacement," *Bulletin of the Seismological Society of America*, **84**: 974–1002.

Wells, W. V. 1857. *Explorations and Adventures in Honduras*. New York: Harper and Brothers, Publishers.

Wernicke, B., G. J. Axen, and J. K. Snow. 1988a. "Basin and range extensional tectonics at the latitude of Las Vegas, Nevada," *Geological Society of America Bulletin*, **100**: 1738–1757.

Wernicke, B., J. D. Walker, and K. V. Hodges. 1988b. "Field guide to the northern part of Tucki Mountain fault system, Death Valley region, California," in *This Extended Land, Geological Journeys in the Southern Basin and Range*, Geological Society of America, Cordilleran Section, Field Trip Guidebook, 58–65.

Wesnousky, S. G. and L. Leffler. 1992. "The repeat time of the 1811 and 1812 New Madrid earthquakes: a geological perspective," *Bulletin of the Seismological Society of America*, **82**: 1756–1785.

White, R. A. 1985. "The Guatemala earthquake of 1816 on the Chixoy–Polochic fault," *Bulletin of the Seismological Society of America*, **75**: 455–473.

White, R. A. and D. H. Harlow. 1993. "Destructive upper-crustal earthquakes of Central America since 1900," *Bulletin of the Seismological Society of America*, **83**: 1115–1142.

Wiegel, R. L. 1970. *Earthquake Engineering*. Englewood Cliffs, NJ: Prentice Hall.

Wiggins-Grandison, M. D. 2001. "Preliminary results from the new Jamaica seismograph network," *Seismological Research Letters*, **72**: 525–537.

Wightman, A. M. 1990. *Indigenous Migration and Social Change*. Durham: Duke University Press.

Willey, G. R. 1966. *An Introduction to American Archaeology*, Vol. 1, *North and Middle America*. Englewood Cliffs, NJ: Prentice-Hall.

1973. "The Altar de Sacrificios excavations," *Papers of the Peabody Museum of Archaeology and Ethnology, Harvard University*, **64**(3): 1–85.

1987. *Essays in Maya Archaeology*. Albuquerque: University of New Mexico Press.

Willey, G. R. and D. B. Shimkin. 1971. "The collapse of Classic Maya civilization in the southern lowlands," *Southwest Journal of Anthropology*, **27**: 1–18.

Willey, G. R. and A. L. Smith. 1969. "The ruins of Altar de Sacrificios, Department of Petén, Guatemala: an introduction," *Papers of the Peabody Museum of Archaeology and Ethnology, Harvard University*, **62**(1): 1–49.

Willey, G. R., A. L. Smith, G. Tourtellot, III, and I. Graham. 1975. "Excavations at Seibal, introduction: the site and its setting," *Memoirs of the Peabody Museum of Archaeology and Ethnology, Harvard University*, **13**(1): 1–56.

Williams, H. 1962. *The Ancient Volcanoes of Oregon*. Eugene: University of Oregon Press.

Williams, R. 1863. "Letter to John Winthrop [1638]." *Collections of the Massachusetts Historical Society*, 4th Series, **6**: 229–230.

Williams, J. S. and M. L. Tapper. 1953. "Earthquake history of Utah, 1850–1949," *Bulletin of the Seismological Society of America*, **43**: 191–218.

Willis, B. 1929. *Earthquake Conditions in Chile*. Carnegie Institution of Washington Publication 382. Washington, DC.

Wilson, D. J. 1988. *Prehispanic Settlement Patterns in the Lower Santa Valley Peru*. Washington, DC: Smithsonian Institution Press.

Winkler, L. 1992. "Catalog of earthquakes felt in the eastern United States megapolis (1850–1930)," *Bulletin of the Seismological Society of America*, **72**: 2285–2306.

Wisdom, C. 1940. *The Chorti Indians of Guatemala*. Chicago: University of Chicago Press.

Witkind, I. J., W. B. Myers, J. B. Hadley, W. Hamilton, and G. D. Fraser. 1962. "Geologic features of the earthquake at Hebgen Lake, Montana, August 17, 1959," *Bulletin of the Seismological Society of America*, **52**: 163–180.

Wolf, E. 1959. *Sons of the Shaking Earth*. Chicago: University of Chicago Press.

Wolters, B. 1986. "Seismicity and tectonics of southern Central America and adjacent regions with special attention to the surroundings of Panama," *Tectonophysics*, **128**: 21–46.

Wood, H. O. 1916. "California earthquakes: a synthetic study of recorded shocks," *Bulletin of the Seismological Society of America*, **6**: 55–180.

1933. "Note on the Long Beach earthquake," *Science*, **78**: 281–282.

1945. "A note on the Charleston earthquakes," *Bulletin of the Seismological Society of America*, **35**: 49–56.

1955. "The 1857 earthquake in California," *Bulletin of the Seismological Society of America*, **45**: 47–67.

Wood, H. O. and F. Neumann. 1931. "Modified Mercalli Intensity Scale of 1931," *Bulletin of the Seismological Society of America*, **21**: 277–283.

Wood, H. O., M. W. Allen, and N. H. Heck. 1939. *Earthquake History of the United States, Part II: California and Western Nevada*, US Department of Commerce Coast and Geodetic Survey 609.

Wood, R. M. 1987. *Earthquakes and Volcanoes: Causes, Effects and Predictions.* New York: Weidenfeld and Nicolson.

Woollard, G. D. 1958. "Areas of tectonic activity in the United States as indicated by earthquake epicenters," *Transactions of the American Geophysical Union*, **39**: 1135–1150.

Working Group on California Earthquake Probabilities (WGCEP). 1990. *Probabilities of Large Earthquakes Occurring in the San Francisco Bay Region, California*. US Geological Survey Circular 1053.

Wright, C. and A. Mella. 1963. "Modifications to the soil pattern of south-central Chile resulting from seismic and associated phenomena during the period May to August 1960," *Bulletin of the Seismological Society of America*, **53**: 1367–1402.

Wright, R. 1989. *Time among the Maya*. New York: Weidenfeld and Nicolson.

Wyss, M. 1978. "Sea level changes before large earthquakes," *Earthquake Information Bulletin*, **10**(5): 165–168.

Yamaguchi, R. and T. Odaka. 1974. "Field study of the Izu-Hanto-oki earthquake of 1974," *Special Bulletin of the Earthquake Research Institute University of Tokyo*, **14**: 241–255.

Yeats, R. S., K. Sieh, and C. R. Allen. 1997. *The Geology of Earthquakes*. Oxford: Oxford University Press.

Yepes, H., J. Chatelain, B. Guillier, *et al.*, 1996. "The $M_w$ 6.8 Macas earthquake in the sub-Andean zone of Ecuador, October 3, 1995," *Seismological Research Letters*, **67**: 27–29.

Yerkes, R. F. and R. O. Castle. 1976. "Seismicity and faulting attributable to fluid extraction," *Engineering Geology*, **10**: 151–167.

Yong, C., K. Tsoi, C. Feibi, *et al.* (eds.). 1988. *The Great Tangshan Earthquake of 1976.* Oxford: Pergamon Press.

Youngs, R. R. and K. J. Coppersmith. 1985. "Implications of fault slip rates and earthquake recurrence models to probabilistic seismic hazard estimates," *Bulletin of the Seismological Society of America*, **75**: 939–964.

Zátopek, A. 1968. "The Skopje earthquake of 26 July 1963 and the seismicity of Macedonia," in *The Skopje Earthquake*. Paris: UNESCO.

Zilbermann de Lujan, C. 1987. *Aspectos socioeconómicos del traslado de la ciudad de Guatemala (1773–1783)*. Guatemala City: Academia de Geografía e Historia de Guatemala.

Zoback, M. D., R. M. Hamilton, A. J. Crone, *et al.*, 1980. "Recurrent intraplate tectonism in the New Madrid seismic zone," *Science*, **209**: 971–976.

# Index